T0299309

Organic Versus Conventional Farming

This book presents the results of a comparison of the quality of food products and raw materials, such as vegetables, fruits and honey, produced in organic and conventional farming systems. The comparison, which was based on literature data and the results of our own research, included not only the chemical parameters important for assessing the nutritional and health-promoting values, but also the stability of the ingredients in the fruits, which is important from a food storage perspective. The ecological, social and economic aspects of organic food production, which are crucial from the perspective of sustainable development, are also discussed in the book.

Organic Versus Conventional Farming aims at academics and farmers, but also to anyone looking for the answer to the question of whether organic farming ensures high-quality food, the production of which is safe for the environment.

Innovations in Environmental Engineering

Editor-in-chief: Lucjan Pawlowski

Associate Editor: Małgorzata Pawłowska

Editorial Board: Kazimierz Banasik, Marzenna Dudzińska, Marek Gromiec, Katarzyna Juda-Rezler, Ewa Klimiuk, Serhii Kvaterniuk, Katarzyna Majewska-Nowak, Hanna Obarska-Pempkowiak, Sergio Orlandi, Artur Pawłowski, Czesława Rosik-Dulewska, Andrei V. Sandu, Corrado Sarzanini, Vladimir S. Soldatov, Christopher G. Uchrin, Maria Wacławek, Tomasz Winnicki, Maria Włodarczyk-Makuła, Irena Wojnowska-Baryła, Guomo Zhou

ABOUT THE SERIES

The 'Innovations in Environmental Engineering' series is devoted to the publication of monographs that aim to integrate environmental engineering and the concept of sustainable development, by counteracting negative changes in the environment with technological methods. Books in the series typically cover the following topics: chemical/biological treatment of wastewater, municipal and industrial waste disposal and recycling, water resources management, improving indoor and outdoor air quality, renewable energy production, life cycle assessment, and risk analysis. Particular attention is paid to the selection of new engineering methods enabling the prevention and neutralization of pollutants in particular parts of the environment: air, water, soil, and modification of production process in the direction of reducing the emission of pollutants into the environment.

BOOKS PUBLISHED IN THE SERIES

Volume 1: The Vehicle Diesel Engine Start-up Process. Operational and Environmental Aspects, 2023, Paweł Droździel

Volume 2: Application of Waste Materials in Lightweight Aggregates, 2024, Małgorzata Franus

Volume 3: Organic Versus Conventional Farming. Nutritional Value and Health Safety of Food Products, 2024, Cezary A. Kwiatkowski, Elżbieta Harasim, Lucjan Pawłowski, Artur Pawłowski, Małgorzata Pawłowska, Barbara Kołodziej

Organic Versus Conventional Farming

Nutritional Value and Health Safety of Food Products

Cezary A. Kwiatkowski, Elżbieta Harasim, Lucjan Pawłowski, Artur Pawłowski, Małgorzata Pawłowska & Barbara Kołodziej

Routledge
Taylor & Francis Group

LONDON AND NEW YORK

First published 2023
by Routledge
4 Park Square, Milton Park, Abingdon, Oxon OX14 4RN

and by Routledge
605 Third Avenue, New York, NY 10158
e-mail: enquiries@taylorandfrancis.com
www.routledge.com – www.taylorandfrancis.com

Routledge is an imprint of the Taylor & Francis Group, an informa business

© 2024 Cezary A. Kwiatkowski, Elżbieta Harasim, Lucjan Pawłowski, Artur Pawłowski, Małgorzata Pawłowska and Barbara Kołodziej

British Library Cataloguing-in-Publication Data
A catalogue record for this book is available from the British Library

Library of Congress Cataloging-in-Publication Data
A catalog record has been requested for this book

ISBN: 978-1-032-46251-6 (hbk)
ISBN: 978-1-032-46253-0 (pbk)
ISBN: 978-1-003-38077-1 (ebk)

DOI: 10.1201/9781003380771

Typeset in Times New Roman
by MPS Limited, Chennai, India

Preface

Organic farming is a multidimensional approach to the production of safe and high-quality food, which takes into account the protection of the natural environment and respects the needs of not only present, but also future generations. The responsible use of natural resources and energy, the preservation of biodiversity and regional ecological balance, as well as care for maintaining the quality of water and soil are the fundamentals of this farming system [European Union Official Website]. Thus, it contributes to sustainable development.

"Organic farming" is defined as production of plant and animal products, without the use synthetic fertilizers, pesticides, transgenic species, antibiotics, and others chemicals supporting growth of the livestock [Francis, 2005]. The term "organic farming" is synonymous with "biodynamic farming", but the latter is related to the idea of the Austrian philosopher Rudolf Steiner, which has been developed since 1924, and is considered a precursor of organic agriculture. It assumes that a "farm" is "an organism" [Paull, 2011]. Biodynamic farming treats the soil as a living system, in which all the factors should maintain life into balance [Harwood, 2018]. The term "organic farming" appeared several years later in the book "*Look to the Land*" by Lord Northbourne [Paull, 2011].

The implementation of the postulates of organic farming should be a guarantee for the consumers that the food purchased by them meets the restrictive requirements at the level of national and EU law. Actions taken by organic farmers and food producers should follow the growing needs and expectations of consumers in the field of healthy food purchase, and can be signed with the European quality mark "Euroleaf".

In this monograph, a critical review of the literature on organic and conventional food quality has been made, and a comparative analysis of selected food raw materials produced in an organic farming system – without the use of chemicals – and in a conventional farming system, has been made. Six popular plant-originated products or raw materials for food production, such as: millet groats, canned green peas, edible carrot roots, zucchini fruits, apples and bananas were subjected to a qualitative analysis. In addition, honey was included into the analysis as a product commonly recognized as having pro-health and antimicrobial properties, being a source of vitamins, as well as supporting the body's resistance in states of fatigue and exhaustion.

It was hypothesized that products and raw materials produced in organic farms are characterized by more favorable nutritional and health-promoting parameters than those from conventional farms. At the same time, it was assumed that a conventional and, in particular, a sustainable farming system may have a positive impact on the quality of agricultural crops and food products in the event of compliance with certain standards related to food production and processing, and allows the production of foods that meet quality requirements.

Author Biographies

Cezary Andrzej Kwiatkowski, Professor, Ph.D., Sc. D. is a researcher and lecturer working at the Faculty of Agrobioengineering of the University of Life Sciences in Lublin. In 2018, he was awarded the title of professor of agricultural sciences. The research interests of prof. Cezary A. Kwiatkowski, mainly include the following issues: optimization of agronomic practices, agricultural systems, improving agrotechnics of herbal plants and vegetables, the effect of biostimulators and other foliar sprays on crop yield and quality, evaluation of the suitability of spent mushroom substrate (SMS) for crop fertilization, aspects of organic farming in agroecosystems, agrotechnical factors influencing soil quality, the role of agriculture in climate change mitigation and multifunctionality of rural areas.

A measurable result of his research is the authorship or co-authorship of 235 papers, including 145 articles in scientific journals, 11 monographs, 24 chapters in monographs and 17 popular science articles, as well as 2 publications related to agricultural practice. The other publications are conference reports. Prof. Cezary A. Kwiatkowski participated in the implementation of 17 research projects, including the manager of 6 scientific grants (2 international) concerning mainly: improvement of agrotechnics of cereal, root and herbal plants, ecological and conventional farming system and sustainable management of mineral and organic fertilizers. Since 2011, he has been the head of the Department of Agritourism and Rural Development operating within the Department of Herbology and Plant Cultivation Techniques.

Elżbieta Harasim, PhD., Sc. D. University Professor is a researcher and lecturer at the Department of Herbology and Plant Cultivation Techniques at the University of Life Sciences in Lublin. The main rescarch problems of prof. Elżbieta Harasim include, among others: improving agrotechnics of cereal, herbal and vegetable plants, the impact of the cultivation system (ecological, conventional) on biodiversity, quality of soil and agricultural crops, the effect of biostimulants and foliar fertilizers on crop yield and its quality, the use of alternative sources of fertilization (biochar, zeolite, digestate, fertilizer whey, mushroom substrate) in plant cultivation, methods of weed control, profitability and economic efficiency in plant production, as well as the role of agriculture in mitigating climate change and multifunctionality of rural areas. Her achievements include over 100 scientific and review publications in peer-reviewed international and national journals as well as 7 monographs and 7 chapters in monographs in renowned book publishers and conference reports. Prof. Elżbieta

Harasim participated in the implementation of 7 research projects, including two international and national agri-environmental programs. She is a member of the Polish Society of Agronomy.

Lucjan Pawłowski, Ph.D., Sc. D., professor of Lublin University of Technology, Poland, Member of the Polish Academy of Sciences and Member of the European Academy of Sciences and Arts. He was born in Poland, 1946. He studied chemistry at the Maria Curie-Skłodowska University. He received his Ph.D. in 1976, and D.Sc. (habilitation) in 1980 both at the Wrocław University of Technology. In 1986 the President of Poland nominated L. Pawłowski as a full professor. He started research on application of ion exchange for water and wastewater treatment. As a result, together with B. Bolto from CSIRO Australia, he has published book "Wastewater Treatment by Ion Exchange" in which they summarised their own results and experience of ion exchange area. In 1980 L. Pawłowski was elected President of International Committee "Chemistry for Protection Environment". He was the Chairman of the Environmental Chemistry Division of the Polish Chemical Society in 1980–1984. In 1994 he was elected the Deputy President of the Polish Chemical Society and in the same year, the Deputy President of the Presidium Polish Academy of Science Committee "Men and Biosphere". In 1999, he was elected as a President of the Committee "Environmental Engineering" of Polish Academy of Science. He is member of editorial board of international journals: "Reactive Polymers", "The Science of the Total Environmental", "Environment International", and "Journal of Ecological Chemistry". In 1991 he was elected as the Deputy Rector of the Lublin University of Technology, and he held this post for two terms (1991–1996). He has published 19 books, over 128 papers, and authored 68 patents.

Artur Pawłowski Ph.D., Sc. D., researcher and lecturer at the Faculty of Environmental Engineering of Lublin University of Technology, Poland. He received his M.Sc. in the philosophy of nature and protection of the environment at the Catholic University of Lublin in 1993. In 1993 he received Ph.D. and in 2009 obtained post-doctoral degree, both at The Cardinal Stefan Wyszynski University in Warsaw, Poland. In 2017, he was awarded the title of professor of technical sciences. He is the Head of the Sustainable Development Unit in Institute of Renewable Sources of Energy Engineering at the Lublin University of Technology.

The scientific interests of prof. Artur Pawłowski focus mainly on multi-dimensional nature of sustainable development, factors connected with climate change, and energy sector, especially in the context of renewable sources of energy. He is Editor-in-chief of the journal Problemy Ekorozwoju/ Problems of Sustainable Development. He authorer or co-authorer more than 200 scientific papers. He is the author of the book "Sustainable Development as a Civilizational Revolution. A Multidisciplinary Approach to the Challenges of the 21st Century (2011) and co-author of the book Biomass for Biofuels (2016). Prof. Artur Pawłowski is a teacher of the international network called The Baltic University, Sweden. He is also a member of the European Academy of Science and Arts in Salzburg, Austria, member of Research Committee Environment, Climate and Energy of the European Academy of Science and Arts, and member of International Association for Environmental Philosophy, Philadelphia, USA.

Małgorzata Pawłowska Ph.D., Sc.D. researcher and lecturer at the Faculty of Environmental Engineering of Lublin University of Technology. She received her M.Sc. in the philosophy of nature and the protection of the environment at the Catholic University of Lublin in 1993. In 1999, she received Ph.D. in agrophysics at the Institute of Agrophysics of the Polish Academy of Sciences, and in 2010 she obtained a postdoctoral degree in the technical sciences in the field of environmental engineering at the Wrocław University of Technology. In 2018, she was awarded the title of professor of technical sciences.

The scientific interests of prof. Małgorzata Pawłowska focus mainly on the issues related to the reduction the greenhouse gases concentrations in the atmosphere, energy recovery of organic waste, and the possibility of using the waste from the energy sector in the reclamation of degraded land. The measurable outcomes of her research include the authorship or co-authorship of 105 papers, including 40 articles in scientific journals, 4 monographs, 24 chapters in monographs, co-editor of 5 monographs, co-authorship of 15 patents and dozens of patent applications. Prof. Małgorzata Pawłowska has participated in the implementation of 9 research projects concerning, primarily, the prevention of pollutant emissions from landfills and the implementation of sustainable waste management. In the years 2013–2019 she was the head of the Department of Alternative Fuels Engineering at the Institute of Renewable Energy Sources Engineering. Currently, she heads the Department of Biomass and Waste Conversion into Biofuels.

Barbara Kołodziej, PhD., Sc. D. is working as a professor and head of Department of Industrial and Medicinal Plants and Laboratory of Quality Assessment of Herbal Raw Materials at University of Life Sciences in Lublin, Poland. Currently, she is Dean of the Faculty of Agrobioengineering at University of Life Sciences in Lublin. She has more than 33 years of teaching and research experience in the fields of biological and agrotechnical aspects of herb and industrial plant production, phytochemistry and bioactive studies. She has been involved in acclimatization of American ginseng in Poland and introduction to the field cultivation of some species that have been extracted from the wild (including protected ones). An important focus of her research is also on the issues related to the phytochemistry of spices and medicinal plants and the introduction of technologies for the management of municipal wastewater through the cultivation of energy crops. Professor Barbara Kołodziej has participated in the implementation of several research projects. Moreover, she has published more than 200 research and review publications in peer-reviewed international and national journals and several books, several dozen of monographs and chapters in monographs in reputed book publishers. She is a member of the Polish Herbal Committee and the Polish Agronomic Society.

Chapter 1

Agriculture and sustainable development

1.1 INTRODUCTION

People can survive in polluted environment, paying with their health and lowered life expectancy. However, they cannot survive without water and food. That is why proper agriculture is one of the key elements of the future of our civilisation and sustainable development.

The sustainable development idea was introduced by the UN in 1987 in report "Our common future". It was defined as "development that meets the needs of the present, without compromising the abilities of future generations to meet their own needs" [WCED 1987]. It is obvious that availability of food should be mentioned among the basic human needs [Millennium Ecosystem Assessment 2005]. There are 8 billion people on our planet. Unfortunately, 1 out of 10 people suffers from chronic hunger and as many as 1 out of 3 people have no regular access to food [UN 2015]. To feed growing population, during next forty years production of food must increase by around 60% [Popp et al., 2014]

In this paper we look through the main challenges for agriculture and possible solutions indicated by the sustainable development programme, including environmental engineering and organic farming.

1.2 AGRICULTURE PAST AND PRESENT

1.2.1 *Warning from the past: the Sumerians*

Since the agricultural revolution, which started 9000 years ago in Asia, people have begun to cultivate chosen plants and breed animals. For centuries such farming has been developing generally steadily.

However, there were also warnings. The earliest advanced civilization of the Sumerians, which flourished in Mesopotamia about 3000 B.C. may serve as an example [Ponting 1993]. It was famous for an invention of a pictographic system of writing [Young 1971]. Their land laid between rivers Tigris and Euphrates, with fertile land, very good for developing agriculture. Through – innovative at that time – irrigation, the yields were stable and high. Availability of food of good quality allowed the population to grow. This also meant a growing demand for food. For a few centuries, the amount of food needed was assured. However, wide use of irrigation had its price: the groundwater level was rising and one of the consequences was that water was taking the salt from deeper layers of soil up to the surface. This meant increasing soil salinity – one of the most dangerous forms of soil degradation. With growing population, they needed

DOI: 10.1201/9781003380771-1

more land for agriculture. Thus, they decided to cut down the forests. In this case, it led to soil erosion – another major form of soil degradation. They were not only losing valuable soil, but it also caused that the rivers were silted up, which led to floods. These processes begun about 2400 B.C. and with passing years they were becoming more intense. About 1800 B.C., the yields dropped by as much as two-thirds, in relation to the best period. Unfortunately, the apparent threats were ignored, which finally led to the collapse of the whole civilization [Ponting 1993].

The catastrophe in Mesopotamia was local. Unfortunately, from the beginning of the 19th century till present, the development of agriculture was responsible for increasing damage of different environments on the global scale [Summers 1996]. At the same time, the average amount of food available was increasing. It is possible due to modern intensive agricultural methods, but it cannot last forever. Loss of soil humus means that the soil may become not good enough to support cultivation.

1.2.2 *Factory farming*

Agriculture will always have an important influence on the environment. Although traditional farming may be more balanced, the most controversial issues are connected with using at least some chemistry, erosion and energy management. In many areas, however, such farming is endangered or almost non-existent. It is so, because in recent decades the market was taken by the factory farming. In such farming, animal breeding and plant cultivation occur on a massive scale. The produced food is very cheap, but there are important environmental and health problems.

From environmental point of view, we must notice huge amounts of waste, like manure. In traditional farming, it may be used as fertilizer for plant crops. However, in industrial farm, connected with breeding of thousands of animals, there are no crops. Even if there were, it is almost impossible to imagine how large the area was supposed to be cultivated to deal with the manure from so many animals. Let us presume that we have a herd of 2500 pigs. It would produce 100 000 000 dm^3 of excrements and 80 000 000 dm^3 of mud. Such 'manure' may contain even 100% more pathogens than it may be found in human excrement. Its storage is difficult and may result in pollution of soil and groundwater [America's Animal Factories 1998; Public Citizen 2004].

Factory farming is also dangerous from the perspective of health. As an example, let us mention the year 2007, when, because of the violation of sanitary procedures and standards on Romanian factory farms, the swine fever epidemic broke out. As a result, 1/3 of the farms had to be closed. The problem returned in the following years, for example in 2018 [Pig Progress 2018]. Health aspects are also connected with huge animal density on such farms, for example chickens crammed in cages. Viruses and bacteria easily spread under such conditions. Preventive actions include massive use of antibiotics served together with the fodder. It leads to immunity of some bacteria to antibiotics, which is a threat not only to animals, but also to the people eating such meat.

Unfortunately, in the USA as much as 99% of the animals bred for food are coming from factory farming [PETA 2022]. Such large production means that most of the traditional farms were liquidated or taken over by factory farms.

What is interesting, in many cases it is not easy to spot the factory farming products, because such companies are buying different brands of food producers to sell the factory produced food under different than their own brands. In Washington, a control in a supermarket revealed that even though articles produced by 50 companies were found, they were owned by one large consortium [Civil Action 2007].

What is the solution? Better education showing that cheap products are not necessarily healthy. Introduction of the legislation that will protect consumers better and force owners of factory farms to increase their care about animals as well as increase their environmental standards. Promotion of sustainable and organic farming and financial support for such farmers.

1.3 THE PERSPECTIVE OF SUSTAINABLE DEVELOPMENT: ECOLOGICAL AND ORGANIC FARMING

As it was already mentioned, the sustainable development idea is about fulfilment of basic human needs. There are 3 basic pillars of sustainability: environmental, economic and social. In the case of agriculture, the most obvious in environmental and premise that environmental footprint should be the lowest possible. However, social and ecological issues are equally important. Agriculture is managed by people and the conditions of their work also matter. Finally, it gives people economic profit [Benkebila 2015; Esnouf *et al.*, 2013].

There are also 17 United Nations Sustainable development goals [Table 1.1].

Table 1.1 Sustainable development goals [UN 2015].

No.	Content
1	No poverty
2	Zero hunger
3	Good health and wellbeing
4	Quality education
5	Gender equality
6	Clean water and sanitation
7	Affordable and clean energy
8	Decent work and economic growth
9	Industry, innovation and infrastructure
10	Reduced inequalities
11	Sustainable cities and communities
12	Responsible consumption and production
13	Climate action
14	Life under water
15	Life on land
16	Peace, justice and strong institutions
17	International partnership for the goals

The most important, from the perspective of agriculture, are [UN 2015]:

– Goal 2: Zero hunger, so a matter of availability of food for the whole human population.
– Goal 3: Good health and well-being, so not only the amount of food, but also its quality matters.
– Goal 7: Affordable and clean energy, farming does not only use energy, but may be its producer, for example in the case of biogas or so called agrophotovoltaics. This area is very closely connected with environmental engineering.

– Goal 8: Decent work and economic growth. Agriculture is not industry, but any agricultural product has its price and its production needs important human resources.
– Goal 12: Responsible consumption and production, which does not only refer to industry, but also agriculture.
– Goal 13: Climate action, so lowering emissions of greenhouse gases from agriculture to the atmosphere.
– Goal 15. Life on land, also agricultural land, where quality of water, air and soil matters.

Some other goals have indirect connection with agriculture. For example, goal 6: Clean water and sanitation. No matter how ecological a farmer wants to be, if the farm is located in an area with polluted water, air and soil – they cannot produce food of good quality.

Clean agricultural environment is connected with idea of healthy ecosystem. Such ecosystem [de Groot 2011]:

– Includes plants and animals that produce organic matter and simple organisms, which break it down.
– Has a large biodiversity and is capable of self-restoration after external disturbances (resilience),
– As a concept often also includes the concern for human health and good agricultural practices, as humans are seen as part of the ecosystem.

Generally, sustainable farming cannot destroy or pollute the surrounding environment and should be self-reliant [Sobol 1996].

There are many forms of sustainable farming, like permanent farming, or organic farming.

The idea of permanent culture was introduced in 1978 by Bill Mollison and David Holmgren. Permanent agriculture is permanent culture and means "the conscious design and maintenance of agriculturally productive ecosystems which have the diversity, stability and resilience of natural systems (...). Permaculture makes use of observations of natural systems and the wisdom contained in traditional farming and social systems, as well as modern scientific knowledge. While it is based on ecological models, permaculture aims to create cultivated ecologies, designed to produce more human and animal food that is generally found in nature. A sustainable system of agriculture is essential if we are to maintain a permanent and sustainable future" [Sobol 1996].

Currently, the most popular term, promoted in international legislation, is organic farming, which is connected with biological, ecological and biodynamic farming [Woodward 1996].

One of the basic definitions of organic farming was formulated by The United States Department of Agriculture [USDA], in which "organic farming is a production system which avoids or largely excludes the use of synthetically compounded fertilisers, pesticides, growth regulators and livestock feed additives. To the maximum extent feasible, organic farming systems rely on the crop rotation, crop residues, animal manures, legumes, green manures, off-farm organic wastes and aspects of biological pest control to maintain soil productivity and tilth, to supply plant nutrients and to control insects, weeds and other pests. The concept of the soil as a living system, develops the activities of beneficial organisms, is here a base" [USDA 2022].

American organic farming is doing well. Between 2011 and 2021, the area covered by organic croplands in the USA increased by 79%. The share of organic food in overall

American food sales in grocery stores, club stores and supercenters in 2021 was 56% [USDA 2022].

One of the assumptions of organic farming may be questionable – that only natural nutrients may be used. We must remember that both living and dead material consists of chemical compounds and it applies also to fertilizers [Woodward 1996]. The nitrogen and phosphorus are the same both in natural and artificial fertilizers. It is rather a matter of dosage, as artificial nutrients are too often used in too large quantities. It is in compliance with the Liebig's Law of the Minimum, which in the case of agriculture states that "growth is controlled by the scarcest resource and not by the total of resources available" [de Vries 2013]. It means that increasing the application of nutrients after some point does not enhance growth. Growth will continue only in the case we are applying the nutrient which is the most scarce (the limiting factor) or all nutrients which are available.

On the other hand, in the case of agriculture every year we have harvest, when some nutrients leave farms in a form of agricultural products. Thus, in the case of nutrient management, we must remember that "the rate of outflow on nutrients does not exceed the rate if inflow by natural and human processes" [de Vries 2013].

There are many organisations supporting organic farming and formulating important principles and guidance. One of them is International Federation of Organic Agriculture Movements. Their principles are presented in Table 1.2.

These principles represent a holistic approach. The food is supposed not only to be natural and of high-quality, but also the care about natural ecosystem and biodiversity were underlined, as well as the necessity to include wider socio-ecological context of organic farms was formulated.

In this section, it is also worth mentioning the legislation of the European Union.

According of the definition from European Commission, organic farming is "an agricultural method that aims to produce food using natural substances and processes. This means that organic farming tends to have a limited environmental impact as it encourages:

– responsible use of energy and natural resources,
– maintenance of biodiversity,
– preservation of regional ecological balances,
– enhancement of soil fertility,
– maintenance of water quality.

Organic farming rules encourage a high standard of animal welfare and require farmers to meet the specific behavioral needs of animals" [EC 2022].

Additionally, for the years 2023–2027, 10 objectives related to the common agricultural policy [CAP] were formulated:

– "to ensure a fair income for farmers,
– to increase competitiveness,
– to improve the position of farmers in the food chain,
– to take action for climate change,
– to include environmental care,
– to preserve landscapes and biodiversity,
– to support generational renewal,
– to include vibrant rural areas,
– to protect food and health quality,
– to foster knowledge and innovation" [EC 2022a].

Table 1.2 The basic principles of organic farming by International Federation of Organic Agriculture Movements [IFOAM 2022].

No.	Principle
1	To produce food of high nutritional quality in sufficient quantity
2	To work with natural systems, rather than seeking to dominate them
3	To encourage and enhance biological cycles, within the farming system involving micro-organisms, soil flora and fauna, plants and animals
4	To maintain and increase the long-term fertility of soils
5	To use whenever possible renewable resources in locally organised agricultural systems
6	To work as much as possible within a closed system, with regard to organic matter and nutrient elements
7	To give all livestock the conditions of life that allow them to perform all aspects their behaviour
8	To avoid all forms of pollution that may result from agricultural techniques
9	To maintain the genetic diversity of the agricultural system and its surroundings, including the protection of plant and wildlife habitats
10	To allow agricultural producers an adequate return and satisfaction from their work, including a safe working environment
11	To consider the wider social and ecological impact of the farming system

It is worth emphasising that special care should be applied to soil, which is the basis for any farming. The sustainable management of soil means the necessity "to preserve its inherent fertility" [de Vires 2013].

Organic farming in the EU is quite successful. Table 1.3 shows total organic farm areas in different European countries and percentage of change between 2012 and 2020.

In the last 10 years, the land used by organic farming in the EU increased by 66% and now covers about 9 457 886 ha. During that time, sales of products from such farming increased by 128% [EC 2022]. The countries where the percentage of land used for organic agriculture is growing most rapidly are Croatia, Bulgaria, France and Hungary. The only country where this percentage decreases (−22.3%) is Poland. It may be a surprise, since Poland is perceived as an agricultural country. The truth is that in Poland many former farmers merely live in villages, while their work is connected with other companies in nearby cities. There is also a new trend that people are moving from cities to villages, but again only to build their homes, their work is still at nearby cities.

The largest land used for organic farming in Europe is in France, Spain and Italy – reaching above 2 million ha in each of these 3 countries.

Unfortunately, all agriculture cannot be organic. The second UN Sustainable development goal is: no hunger. Almost 1/8 of the world population, mainly in Asia and Africa suffers from hunger. In this case, ANY food is needed.

There is also a problem that usually organic agriculture can deliver smaller yields than the traditional variant. It is true in the countries like the USA, where factory farming dominates the whole market. However, there are poorer countries, like Cambodia, where organic farming gave 71% yield increase [Cooney 2006; Pawłowski 2011]. Some authors believe that organic farming is the future of agriculture and we are witnessing another Agricultural Revolution – this time both Green Revolution and Livestock Revolution [Jaggard *et al.*, 2010; Thornton 2010].

There are factors, however, which can stop this emerging revolution – the threats connected with the climate change and the global warming.

Table 1.3 Total organic farm areas in the EU by country
        [EC 2022b].

No.	Country	Organic area in 2020 in ha	Percent of change between 2012/2020
1	Belgium	99 072	+65.9
2	Bulgaria	116 253	+197.0
3	Czechia	540 375	+15.3
4	Denmark	299 998	+54.1
5	Germany	1 590 962	+65.8
6	Estonia	220 796	+55.4
7	Ireland	74 666	+41.4
8	Greece	534 629	+15.6
9	Spain	2 437 891	+38.8
10	France	2 517 478	+144.2
11	Croatia	108 610	+240.4
12	Italy	2 095 364	+79.5
13	Cyprus	5 918	+50.9
14	Latvia	291 150	+48.8
15	Lithuania	235 471	+50.4
16	Luxembourg	6 118	+48.1
17	Hungary	301 430	+130.8
18	Malta	67	+81.1
19	Netherlands	71 607	+49.1
20	Austria	671 703	+26.0
21	Poland	509 286	−22.3
22	Portugal	319 540	+59.1
23	Romania	468 887	+62.7
24	Slovenia	52 078	+48.4
25	Slovakia	222 896	+35.6
26	Finland	316 248	+59.9
27	Sweden	610 543	+27.0
28	EU-27	9 457 886	+55.6

1.4 CLIMATE CHANGE AND AGRICULTURE

Modern agriculture contributes to the global climate change, since in the case of emissions of the main greenhouse gases even 25% comes from farming [Dickie *et al.*, 2014]. On the other hand, agriculture suffers from climate change, so it is a two-way relation. If the future climate will not be suitable for agriculture, there will be no farming at all and our civilization will end. We can already witness many anomalies with stronger and more frequent tornadoes, thunderstorms, hurricanes, flooding and – at the same time – prolonged droughts in different parts of the world [Gates 2021; Meadows *et al.*, 1992; Rom 2022].

However, even worse may happen: many areas on our planet may be flooded forever. One of the endangered places is the Kiribati country, lying on the Pacific islands, populated by 100 000 people. Because of the rising level of the ocean, the country already lost two islands, totally flooded in 1999. It is estimated that all of the islands will be under water by the end of the 21st century. These which still exist are being penetrated by the ocean water more frequently. Such water is salty and can make local freshwater unusable. It also s destroys not only local crops, but even buildings. It is worth noticing that the people from this region are emitting only 0.6% of the global greenhouse gases responsible for the global warming and raising the ocean level [Iberdola 2020]. Can we help them? Well, there was a project to build a platform floating on the sea, which could replace the sinking island. However, the estimated cost was 2 billion USD, much beyond the financial possibilities of people of Kiribati. In turn, they were able to buy some land on Fiji, where crops are safe and which may also serve as the place for the people who will be forced to evacuate from Kiribati. Because of the climate situation, they were also promised by New Zealand's government that every year 75 citizens of Kiribati will be allowed to immigrate to this country [Iberdola 2020]. It is believed that climate change will be the greatest reason for human migrations by the end of the 21st century. In 2022 alone, 10 largest climate anomalies have made a loss of 170 billion USD and more than 100 million people were refugees because of climate change, accompanied by hunger or war [Kordylas 2022].

However, it does not have to be this way, especially in the context or reduction of emissions of the main greenhouse gas: carbon dioxide. Most of the countries pledged to become carbon neutral by the year 2050, Norway has the most ambitious plan to achieve this goal in 2030 – see Table 1.4.

Table 1.4 Announced dates for gaining carbon neutrality by different countries [Guterres 2020].

No.	Country	Pledged date
1	China	2060
2	European Union	2050
3	Finland	2045
4	Great Britain	2050
5	Japan	2050
6	Norway	2030
7	Republic of Korea	2050
8	Sweden	2045

An important part of this reduction must be connected with agriculture; however, at the same time, we must achieve greater productivity of farms to ensure the food for all people.

There are many initiatives that are trying to reshape the agriculture in these directions, from UN Framework Convention on Climate Change [UN 2006] to the Global Research Alliance on Agricultural Greenhouse Gases [GRAAGG 2022]. They are trying to introduce new guidelines for farmers, but also implement new economic instruments (or example subsides) that will help farmers to meet new standards.

The following section was based on the report "Strategies for Mitigating Climate Change in Agriculture: Recommendations for Philanthropy" [Dickie *et al.*, 2014], covering the time till the end of the year 2030. The goals and possible reductions adopted in this document from the world's perspective are presented in Table 1.5.

Generally, as much as 6.3 Gt of CO_2eq. may be saved.

The first goal is connected with the possibilities of increasing efficiency in agricultural production. The possible reduction of CO_2 is impressive – 2 Gt of CO_2eq. every year. How to achieve this? Let us note that a huge part of the CO_2 emissions from agriculture is connected with livestock breeding, especially cattle. An important solution is related to changes in cow's diet. There are two possible solutions.

The first is reduction of the methane (an important greenhouse gas) emissions of enteric nature, which may be achieved by adding supplements than influence digestion and its microbiology.

The second goal is to prepare a new diet, with the help of which livestock could gain weight faster, so the time of breeding will shorten. At the same time, they are supposed to produce not only more milk, but also more meat.

Table 1.5 Possible CO_2 reductions in agriculture to the year 2030 [Dickie et al., 2014].

No.	Goal and area of agriculture	Possible CO_2 eq. reduction for every year after 2030
1.	Increase in efficiency of agricultural production	2 Gt
2.	Better manure storage	260 Mt
3.	Increase the efficiency of nutrient fertilizers use	325 Mt
4.	Possible carbon sequestration by terrestrial ecosystems	1600 Mt
5.	Necessity to change consumption patterns	3 Gt

In the case of plants, the solution is called intensification. It means production of more food without expanding the area taken by the arable land [Burney et al., 2010]. However, one must be careful here, because intensification may cause biodiversity loss [Mineau 2022].

The second goal is also about cattle. They produce not only milk or meat, but also manure, the storage of which should be safer and better. In this case, we can save further 260 Mt of CO_2eq. every year.

Generally, there are two possibilities:

— Tanks (chambers) where biogas may be collected and burned to produce heat or electricity –this issue will be discussed later.
— Utilization of manure as compost. Here, the already discussed organic farming may play a crucial role, since in such farming there is a deep integration of plant crops and breeding of livestock.
— The third goal is connected with increased efficiency of fertilizer dosing (both synthetic and natural). Here, we are also dealing with another important greenhouse gas – nitrous oxides, and possible reduction of CO_2 equivalents amount to 325 Mt of CO_2eq. every year. It may be done through:
— Strong connection of fertilizer dosing with plants needs, with the use of computers and regular soil testing. Along with it, we will also know what the actual concentration of nutrients in soil is.
— Genetic modifications of plants, which may help to increase the uptake of nutrients from the soil and, at the same time, lower the concentration of nutrients in soil.
— Generally, realisation of this goal should also be connected with more adequate land use, better management of crops and resignation from deforestation.

Stopping deforestation is especially important, because forests stabilize the climate and accumulate water, thus counteracting floods. Deforestation means more frequent and more violent climate anomalies. For example, deforestation was the cause of a huge flood in China in 1998 in the Yangtze valley. Approximately 3700 people died during the flood and in the case of field destruction, they lost 60 million acres. Loses were estimated at 30 billion USD. However, a new programme was soon introduced: no more deforestation in this region and introduction of reforestation at the expense of 12 billion USD [Lovins 2004].

Forests are also endangered in the areas where deforestation is not planned. For example, in California the trees ale literally dying because of climate anomalies: high temperature and lack or water. During the summers of 2020 and 2021 both the snow-pack and precipitation were 40% lower than the average. In 2015, not even a single rainfall event was noticed in this region [Pacific Institute 2022].

The fourth goal is part of a wider issue of possibilities of carbon sequestration by terrestrial ecosystems. In the case of agriculture, 1600 Mt of CO_2eq. may not be emitted every year. Generally, the greatest amounts of carbon dioxide may be absorbed by forests, so afforestation wherever possible is one of the most important factors. In the case of agriculture, orchards, pastures, meadows, cereals and industrially cultivated plants can be mentioned [Pawłowski 2020].

Moreover, about 1 500 Gt of carbon are currently saved in the soil and should stay there. This is the issue of the best protection of the surface of the Earth. If we want to counteract erosion (both wind and air), for example, afforestation is a must. Agriculture may play an important role here, since there is a new trend called agroforestry. It is "the intentional integration of trees or shrubs with crop and animal production to create environmental, economic, and social benefits" [USDA 2019].

There at least five directions in which trees may be involved in agriculture [USDA 2019] as:

1. windbreaks (shelterbelts).
2. riparian forest buffers, especially along waterways.
3. silvopasture systems.
4. forest farming, i.e. farming inside of the forest, for example of herbal or medicinal plants.
5. alley cropping.

The fifth goal is of different nature. Generally, the agriculture gives us the food that we want. However, changing the consumption patterns may reduce all the possible CO_2eq. emissions from agriculture every year by half.

Approximately 75% of it may be achieved by modifications in our diet. Here, reduction of our meat consumption is important, especially beef (which is responsible for 35% of direct CO_2 emissions from agriculture, if we add all ruminants to cattle, this factor will be at the level of 63%). However, if we do not want to resign from eating meat, maybe it is possible to produce it in a different way? There is a new technology, which is allows producing meat from muscle stem cells in vitro [Su et al., 2013].

About 25% of possible CO_2eq. emissions reduction is connected with reduction of waste, especially reduction of food waste.

Another important factor that may improve agriculture, is on the energy side and will be discussed in the next paragraph.

1.5 RENEWABLE SOURCES OF ENERGY AND AGRICULTURE

In many strategies connected with organic farming, wide use of renewables is equally important as right farming procedures. In many countries, farmers may not only be consumers of energy, but also prosumers – people/ farms producing electricity for their own needs with option of selling it to the energy grid [UNDP 2000].

Renewables are very strongly supported in many regions of the world. The climate and energy policy introduced in the European Union is perhaps the best.

In 2009, the first horizon was established and one of the main goals was that 20% of energy of the EU countries by 2020 should be obtained from renewable sources of energy [EWC 2020]. The goal was indeed obtained [EEA 2021]. What is important, similar policies were introduced by other European countries, which are not members of the community. In Iceland, the share of renewable sources of energy has become as high as 83.7% and in Norway 77.4% [Pawłowski 2021]. In the case of the latter, it was mainly hydropower, and in Iceland also geo-thermal sources.

The next horizon, introduced in European Green Deal in 2019, set the goal that by the 2030 at least 32% energy in the Community will came from renewables [Fetting 2020].

A totally new situation was created by the war in Ukraine. The reaction of European Commission was very strong. On May 18th 2022, a new strategy was intro-duced, called "REPowerEU: Joint European action for more affordable, secure and sustainable energy". The document assumed the necessity to end the EU's dependence on Russian gas and oil. It was not only about looking for other suppliers, but also about acceleration of support for renewables. A new target for the 2030 is 45% of energy in the EU from renewables, so 225% more than in 2020 [EC 2022c]. Taking into account that we have the year 2023 already, it may be difficult to fulfill this goal in 7 years, but even if it is not met, huge progress in renewable sources of energy will be visible.

Promotion of renewables in Europe is addressed to all sector of economy, so also to all kinds of agriculture and strongly connected with environmental engineering. There are a few very important problem areas that are worth discussing.

First of all, the so-called agrivoltaics or agrophotovoltaics. The concept was introduced in 1981 by Adolf Goetzberger, together with Armin Zastrow. The main idea is to introduce solar panels into crops. The challenges are, inter alia, to ensure enough light for both factors and that new installations cannot interfere with harvesting devices [Weselek et al., 2019].

Other important factors, in the context of renewables in agriculture, are biomass, biofuels and biogas.

Biomass is peat, straw, wood waste and special plantations. In Europe, willow, poplar and rosa multiflora are popular.

As for now, the largest biomass power station is Drax in North Yorkshire in the United Kingdom. Its capacity is 2 595 MW [Irena 2020].

Biomass has many advantages [Mackow et al., 1993]:

- Fast growth.
- High calorific values.
- Fits to almost every climate, temperature and soil.
- Immunity to most diseases and pests.
- Stability of use and high crops (energy plantation of willow may last for 25 years, and harvests are possible every 3–4 years).

There are also some disadvantages:

- Development of biomass in the EU in years 2007–2010 caused a 2.5-fold increase of the prices for food.
- Converting existing ecosystems to energy plantations is connected with significant carbon dioxide emissions, up to 400 times more, than year's savings from using energy plants. That is why in the EU there is a legislation that only such biofuels that emit now at least 70% (and even 10% more in 2026] less greenhouse gases that are emitted from burning of fossil fuels are allowable [EC 2001].
- If biomass is local, there is no problem, but if there is a need to transport it over longer distances, energy efficiency of biomass is reduced.

Using biomass we should look for low-carbon technologies, controlling the carbon in biomass supply chain and stabilizing land carbon resources [Committee on Climate Change 2018; Daioglou *et al.*, 2019; Miras-Avalos & Baveye 2018; SRU 2007].

The most important is a simple fact that the biomass is almost everywhere. As for now, we are using only 2/5 of the possible biomass resources [Vlosky & Smithart 2001], so there are great possibilities for the future.

Another renewable source of energy which is connected with agriculture, is production of liquid biofuels. The most popular biofuels are bioethanol and biodiesel [Fulton *et al.*, 2004; Tomes *et al.*, 2011].

Bioethanol is produced from sugar-rich plants, i.e. from corn, sugarcane, grasses, willows or even agricultural (for example wheat straws) and municipal (like paper) residues and wastes [Koh & Ghazoul 2008; Pandey *et al.*, 2015].

In the case of bioethanol important source are oilseeds, like rapeseed [Koh & Ghazoul 2008].

There are many new possibilities for biofuels, like plantations connected with the already mentioned agroforestry or aquatic macroalgaes, e.g. green algae [Campbell 2008].

In the European Union, the greatest pressure now is for replacing traditional vehicles with electric ones, since the level of pollution from cars and trucks burning fossil fuels is continually growing [Tengstrom & Thynell 1997]. However, a lot of electricity in the community is still being produced from coal burning plants, so electric cars are not necessarily so ecological, it also depends on the way that this electricity was manufactured. The biofuels from agriculture could be an alternative, especially because the emissions of pollutants from biofuels are much lower, than in the case of traditional gasoline and diesel fuels – see Table 1.6.

Finally, biogas. It may be produced during the fermentation processes in wastewater treatment plants, municipal landfills and also in special biogas plants, which may be connected with agriculture [Burg *et al.*, 2023].

The largest biogas installation was built in Denmark – in Korsko. It deals with 1 million tonnes of agricultural waste annually, and can produce 45.4 MW of electricity [Kumar 2017].

The main elements of biogas are CH_4 and CO_2. Methane may be burned to produce heat and electricity. It may be also used as another kind of biofuel for cars, and such installations are found, for example, in Sweden, the Netherlands and Switzerland.

Table 1.6 Emissions of toxic substances from different biofuels in comparison to fuels produced from oil [Bułkowska *et al.*, 2016].

Fuel	Emissions reduction	Emissions increase
Bioethanol E85	15% reduction of VOC emissions 40% reduction of CO emissions 20% reduction of particulate emissions 10% reduction of NO emissions 80% reduction of sulphur compounds Generally lower emission of hydrocarbons	Increase of emissions of ethanol and acetaldehyde
Biodiesel B20	10% reduction of CO emissions 15% reduction of particulate emissions 20% reduction of sulphur compounds emissions	26% increase of NO_x
Biodiesel B100	50% reduction of CO emissions 70% reduction of particulate emissions 40% reduction of hydrocarbons emissions	9% increase of NO_x emissions

A biogas plant includes gas holder (chamber) and digester. In the latter, the production of biogas takes place through anaerobic digestion [Kumar 2017]. Then, the biogas is collected in the chamber and ready to burn. During the combustion of biogas, there is only a small emission of CO_2. It must be underlined that burning biogas is safe from the perspective of climate change. The already mentioned emissions of CO_2 are balanced by burning of methane – a greenhouse gas with as much as 30 times higher warming potential than carbon dioxide. Additionally, biogas installations are not expensive.

The chapter presents the challenges of contemporary agriculture from the perspective of sustainable development. It is not only a matter of how to ensure the food for 8 billion people, but also how to ensure good quality of this food. The task is difficult, since most of the market was taken by the factory farming. Such food is cheap, but not necessarily healthy. Fortunately, the idea of much better organic farming, producing high quality food and taking care about agricultural environment, has gained popularity in recent years. Some scientists even say that it is the beginning of a new Agricultural Revolution. However, there is a threat – climate change and global warming. Modern agriculture is responsible for about 25% of dangerous greenhouse gases emissions. At the same time, farming is suffering from climate anomalies caused by these emissions. There are solutions however, how to make this emission radically lower, for example by changing cow's diet, intensification of farming or carbon sequestration by terrestrial ecosystems. An important part of the solution is about energy shift. Farming should not only use renewable energy sources, but it can also produce energy, for example by introducing agrophotovoltaics. Agriculture is also important, because it produces biomass, biofuels and biogas.

However, the most important factor is our own choice. Half of the possible carbon dioxide reductions is connected with our behaviour. We decide on our diet or which products to buy. Therefore, it is not only a matter of acquiring right agricultural technologies, but also about our will to make the change. The old slogan "act locally, think globally" is more relevant than ever.

Chapter 2

Farming systems

2.1 INTRODUCTION

The literature uses various definitions of farming system divisions of agricultural production systems and farming systems. Farming production system (farming system) is an economic and agricultural concept that describes a farm holistically (through a set of many variables and indicators) in terms of agricultural land use, i.e. crop and livestock production, the type of non-agricultural economic activities carried out (sources, ways and efficiency of earning outside agriculture), income and life of members of the agricultural household, as well as in terms of the natural, social, economic, infrastructural and institutional environments that determine the listed economic activities. Each farm has its own unique agricultural production system. Typically, there is a wide variety of agricultural production systems, not only at the large scale of geographic space, but also within limited rural areas or targeted types of these systems [Teng & Penning de Vries 1992; Vermuelen *et al.*, 2013]. The name of the main or targeted type of agricultural production systems often includes the name of the agricultural system, which defines the approach to agricultural production in the system under consideration, in terms of the inputs to production, the environmental burden, as well as the degree of environmental and socioeconomic sustainability [Alemayehu *et al.*, 2022; Darnhofer *et al.*, 2010].

An agricultural system or farming system is most often defined as a way of managing agricultural space in terms of plant and animal production and their processing, valued by ecological and economic criteria. In modern agriculture, three farming systems (agricultural systems) are distinguished: conventional, ecological, and integrated system – which is a sustainable, environmentally friendly resultant form of both systems. The basis for distinguishing the systems is the degree of agriculture's dependence on industrial inputs, mainly mineral fertilizers and pesticides, and its impact on the natural environment [Jones *et al.*, 2017].

According to Dixon *et al.* [2014] many factors and development opportunities have influenced the evolution of agricultural systems. Overall, the transition from traditional to modern agriculture, including a series of interrelated sub-trends such as changes in farming methods, crops and varieties, represents a major shift in both farming systems. The components of adaptive capacity have been affected by modernization, resulting in both positive and negative impacts. The most notable variances are primarily related to diversity. Farming systems continue to be significant sources of food, income, and livelihood, despite various changes.

This chapter presents an overview of information on the comparison of different agricultural systems, without explicitly indicating the superiority of any of the systems.

DOI: 10.1201/9781003380771-2

Today, it is important to better understand the adaptability of agricultural systems to changing climatic, economic, technological and social conditions. Thus, it is necessary to consider how to better integrate modern agricultural methods to maintain diversity, rather than a general approach to a particular farming system. In addition, further research is required on how socioeconomic conditions, as well as climate change, may affect yields, food quality, diversity and the natural resource base in rural areas. Modern agriculture should take into account the trade-off between the productivity (economic profit) obtained and the long-term ability to maintain high quality agricultural crops. Decision-making processes (farm management) inside and outside the agricultural system are also important.

2.2 CHARACTERISTICS OF CROP PRODUCTION IN THE ORGANIC SYSTEM

Organic farming currently plays a significant role in striving to produce high-quality food, which is a key tool for building product health security while preserving biodiversity. Authors of studies outlining both the strengths and weaknesses of this type of farming raise questions of environmental performance. In some cases, regardless of the cultivation treatments used, factors related to production, which include the distance of suppliers from farm fields, can dominate the idea of sustainability, which should be the foundation of organic farming. However, this is an argument that applies only to a certain percentage of farms that produce food under the organic system. Despite insignificant drawbacks, this agriculture undeniably counters many abuses in the aspect of nature conservation and food production [Tal 2018].

Organic farming is a holistic system that respects all the needs of plant species, taking into account environmental and soil requirements, as well as the care treatments used. The basis for the functioning of this system is the appropriate selection of plants in a given habitat and crop rotation, taking into account natural barriers to cultivation. The starting point in organic production is organic seed and planting material, which excludes the use of material produced by genetic engineering methods. Care should be taken to ensure that the vegetation is as resistant as possible to environmental stresses, diseases, pests, and that it is characterized by high competitiveness against weeds. Plant-specific features, i.e. the root system of legumes, promote soil fertility. In turn, the above-ground parts of these plants influence weed suppression, thus promoting yield optimization [Chhetri et al., 2013].

Since each organic farm is defined as a separate unit, there is a high degree of variation among individual crops. The main principles that farmers follow in this system include:

- preserving biodiversity in all ecosystems,
- following the idea of sustainability,
- maintaining integrity of plant and animal production,
- providing natural protection from crop pests [Kuepper & Gegner 2004].

The main goal of organic farming is to achieve optimum yields through the use of plant species suitable for the habitat. This ensures that the environmental stress on plants is reduced, the cultivation area is properly managed, and the soil is protected from harmful climatic factors. The soil, as the basis for obtaining valuable food raw materials, is seen in the organic system as a determinant to ensure that the land can be used for food production in the future. Hence, it is essential that it be covered with vegetation for as long as possible during the year, preventing the loss of many valuable components and, consequently, its depletion. The use of artificial pesticides is therefore excluded, with an emphasis on controlling pests mechanically and using their natural predators. Environmental resources are used by organic farmers in a sustainable manner, respecting the food and livelihood needs of future generations [Kwiatkowski & Harasim 2019; Smith *et al.*, 2019].

The preservation of biodiversity is achieved through proper yielding, selecting plant species appropriate to the conditions of the habitat, or planting companion plants. Such grouping of individual varieties allows achieving natural protection by regulating weeds, as well as shielding from wind and sun, and making water available through some species with deeper roots. In turn, crop rotation takes into account the sustainable supply of nitrogen compounds to the soil and the provision of a source of nutrients. Proper care constitutes the basis when growing plants in an organic system. This includes avoiding dangerous chemicals and practices that are harmful to soil organisms, such as excessive tillage [Kuepper & Gegner 2004].

Organic agriculture uses renewable resources in its operation. Organic plant production is characterized by high biological diversity, while protecting natural resources, contributing to increased soil fertility and preventing soil erosion. Plants grown organically are nourished naturally by means of the soil ecosystem, avoiding the addition of artificial fertilizers. The artificial fertilizers used in conventional agriculture contribute to greenhouse gas emissions, which in turn create the greenhouse effect [Muller *et al.*, 2017]. The management of organic production is very difficult, as it requires using the right plant varieties, crop rotation, recycling of organic matter, as well as appropriate cultivation techniques. Livestock production also plays an important role in organic production, as it provides the necessary organic matter and nutrients to the cultivated soil, while improving its condition and making agriculture more sustainable. Establishing an organic system on a farm begins with building soil fertility, which requires a transition period – the so-called conversion. This is the time needed to largely restore the balance of the ecosystem through the breakdown of hazardous substances from previously used fertilizers. This period does not authorize the use of the organic farming label on products, although it is subject to inspection by certification bodies [Tsvetkov *et al.*, 2018; Smith *et al.*, 2019].

It should not be disputed that organic farming contributes to the improvement of the environment, offsetting the changes made under the influence of conventional farming. In addition to the beneficial effects on the environment, society also attributes a health-promoting character to organic food, namely, it is supposed to contribute to maintaining optimal health of the organism, reducing the risk of developing chronic disease entities, due to the presence of large amounts of desirable bioactive substances. Also significant is the minimized uptake of harmful substances by plants, which are common in conventional crops due to the use of synthetic fertilizers and pesticides [Hurtado-Barroso *et al.*, 2019].

Consumers have a full range of organic products available in popular supermarkets, health food stores or organic markets – these types of food are increasingly accessible to shoppers. Consumers' purchasing preferences indicate an increasing interest in organic food, which is related to the perception of its higher quality compared to conventional food. Agricultural producers must meet a number of requirements in order for the resulting product to bear the European organic farming label (Figure 2.1). Consequently, buyer confidence in this type of food is being increased [Ditlevsen *et al.*, 2019].

Figure 2.1 European organic food logo [Annex 5 Regulation Par-liament of the European Parliament and of the Council (EU) 2018/848 of May 30, 2018 on organic production and labeling of organic products and repealing Council Regulation (EC) No 834/2007].

Many publications raise the issue of the increased nutritional value of organic produce relative to conventional food. Due to the preservation of proper soil culture in organic farming, the root systems of plants take up much smaller amounts of harmful substances. Organic food is therefore characterized by much lower amounts of nitrogenous compounds – nitrates and nitrites. Increased accumulation of these substances in the soil is conditioned by intensive management with the use of herbicides, as well as sowing and planting plants in insufficiently exposed areas [Kapusta-Duch *et al.*, 2021]. As the results of a study by Rutkowska [2001] show, a significant reduction in nitrates is possible in organic crop production methods. It was shown that in the potatoes from conventional cultivation, the content of these compounds was up to half that of the organic counterparts.

The results of studies on the content of vitamins, especially vitamin C in organic plant products, are similar. Ecologically grown crops are characterized by a higher content of ascorbic acid than conventional products, which translates into their health-promoting properties, as this compound contributes to the proper functioning of the human immune system. In addition, due to their ability to reduce nitrosamines with an increased supply of vitamin C, organic foods are considered one of the tools for cancer prevention. On the other hand, no significant difference has been noticed between the content of heavy metals, such as cadmium, lead or mercury, in organic and conventional products. This is due to general environmental pollution, which includes the metallurgical industry, the use of transportation vehicles and the use of mineral fertilizers in the conventional system [Cardoso *et al.*, 2011; Rembiałkowska 2007].

Organic products and raw materials are characterized by a number of parameters, the values of which are often higher than those of conventional crops [Ostandie *et al.*, 2021]. Such a state of affairs can be observed by analyzing the results obtained by Achremowicz *et al.* [2016], who showed that the biochemical composition of oatmeal obtained by organic methods is much more favorable than that of conventional oatmeal. The high value of the organic product in this case consists of a higher content of dietary fiber, starch, and total ash, which determine the nutritional value and are therefore valued components. In addition to the chemical composition in favor of organic oatmeal, this product is characterized by a much lower content of contaminants that are dangerous to the proper functioning of the body, such as manganese (Mn) and cadmium (Cd). The results of the analysis pertaining to the content of heavy metals:

cadmium and lead (Pb) in the roots of edible carrots are similar. The above statement is also supported by the results of Shvachko *et al.* [2021] and Zhang *et al.* [2021]. As demonstrated by Tońska *et al.* [2017], both elements were present in higher amounts in conventional raw material. Plants are exposed to the uptake of heavy metals through the soil – via the root system, from where they reach other organs of the plant that accumulate elements. The concentration of elements depends on the morphological structure, species characteristics or maturity of the plant [Czeczot & Majewska 2010].

The reason for the presence of bioactive substances of value to health in organic crops in increased quantities is that they take up significantly less nitrogen compounds than plants grown in a conventional system. The so-called carbon-nitrogen theory states that plants in organic agriculture, due to the lack of excess nitrogen uptake, generate processes leading to the production of compounds containing carbon atoms in their structure. This is how the substances useful in the quality assessment, especially those of an antioxidant nature, are produced. Organic products, in addition to a greater amount of vitamin C, are also characterized by a higher proportion of polyphenols. In addition, due to the fertilization of the soil organically and lack of chemical fertilizers, organically grown plants absorb greater amounts of metabolically useful compounds with a lower proportion of undesirable substances and pollutants [Christaki *et al.*, 2012; Zarzyńska *et al.*, 2016].

In contrast, a study by Dangour *et al.* [2009] found that there are non-negligible differences in nutrient content between organically and conventionally produced foods. While these content differences are biologically plausible, they are unlikely to be of public health significance. The conclusion that can be drawn from this review is that there is no evidence to support the choice of organically produced foods over conventionally produced foods to increase intake of specific nutrients or nutritionally relevant substances. The authors believe that further research is needed in this area for a better understanding of the various factors (beyond the production regime) that determine the nutrient content of organic and conventional foods.

Consumer concerns about the possible adverse health effects of food produced by intensive farming methods have led to considerable interest in the health benefits of organic crops and organic animal products. There seems to be a widespread belief among consumers that organic methods result in food with higher nutritional value. It should be noted that there is no conclusive evidence in the scientific literature to confirm or refute this perception. The most consistent findings are for higher nitrate and lower vitamin C content in conventionally produced vegetables, especially leafy vegetables. Information on the possible effects of diets containing organic or conventional produce on human and animal health is scarce. Data from controlled studies in animal models, especially within a single species, are limited or poorly designed, and the results of these studies provide conflicting conclusions. There are no reports of controlled intervention studies in humans in the literature. Comparisons of health outcomes in populations that habitually consume organically or conventionally produced foods are flawed because of the large number of confounding factors that may contribute to the differences that occur [Williams 2002].

The composition of nutrients and other nutritionally relevant substances in all natural products can differ for various reasons, including the production methods used, espacially those which regulate the use of chemical fertilizer, herbicides, and pesticides. Certified organic regimens require strict control over the use of chemicals and medicines in the production of food [Food Standards Agency 2002]. Lairon [2010] points out that organic agriculture developed to date has the potential to produce quality products with some significant improvements in terms of antioxidant phytomicronutrient content, nitrate accumulation in vegetables and levels of toxic phytochemical residues. The author concludes that organic vegetable products tend to have more dry matter, more of certain minerals (Fe, Mg) and antioxidant micronutrients (phenols, resveratrol), while organic animal products have more polyunsaturated fatty acids. As for safety issues, the vast majority (94–100%) of organic foods contain no pesticide residues, organic vegetables contain significantly less nitrates (about half), and organic grains generally contain comparable levels of mycotoxins to conventional ones.

Vigar *et al.* [2019] note that observational studies are producing a growing number of important results linking apparent health benefits to the level of organic food consumption. Studies on pesticide excretion have shown a reduction in urinary metabolites of pesticides with an organic diet; however, there is insufficient evidence to demonstrate a translation into clinically relevant and meaningful health outcomes. If measurable health benefits with organic foods are to be studied, there is a need for research beyond simply measuring reductions in pesticide exposure. The observation that eating organic food significantly reduces urinary OP levels is important information for the consumers who would like to minimize pesticide exposure. Given current knowledge of the toxicity of these chemicals, it can be anticipated that this will translate into health benefits for humans and animals.

Popa *et al.* [2019] also argue that despite too few studies to date and limited evidence, organic food products appear to have higher nutritional value compared to conventional products. Other reasons in favor of consuming organic food include less exposure to antibiotic-resistant bacteria or activation of the natural defense system against pesticides. Positive results in several other areas have also been reported in connection with the consumption of organic products, such as reduced incidence of metabolic syndrome, high BMI, non-Hodgkin's lymphoma, infertility, birth defects, allergic hypersensitivity, otitis media and pre-eclampsia. However, current evidence does not allow a definitive statement on the long-term health benefits of an organic diet. Consumption of organic foods is often associated with overall healthier eating practices as well as lower levels of overweight and obesity, which is likely to influence the results of observational studies.

2.3 CHARACTERISTICS OF CROP PRODUCTION IN CONVENTIONAL AND INTEGRATED SYSTEMS

The conventional system of agricultural production (intensive, industrialized, high-input) is a method of farming aimed mainly at maximizing profit, achieved through high plant and animal productivity. The development of conventional agriculture in Europe took place in the second half of the 20th century. In conventional agriculture, which uses production technologies based on high consumption of industrial inputs (mineral fertilizers, chemical pesticides) and very low labor inputs, great emphasis is placed on continuous improvement of chemical and technical inputs. This has resulted in, among other things, increasing yields and improving productivity in agriculture. Simultaneously, as production intensity increased, there was an improvement in labor productivity, which led to a significant reduction in the number of people making a living from agriculture. The area of farms increased along with production [Alix and Capri 2018; Sahm *et al.*, 2013].

Conventional agricultural producers, in order to achieve their goal, have made the following improvements in crop food production over the years:

- increasing the acreage under cultivation,
- using biological markers,

- introducing in vitro methods to shorten the breeding cycle,
- using significant amounts of chemicals in agriculture,
- using genetic engineering techniques (genetically modified organisms),
- using state-of-the-art machinery, equipment and soil irrigation systems [Sumberg & Giller 2022].

In addition to the benefits of conventional agriculture, providing societies with large inputs of economically produced food, it is also the cause of many environmental risks. The intensification of agriculture is associated with the widespread problem of contamination of the soil and other environmental components: water and air. Hazardous substances that enter circulation contribute to the contamination of food with heavy metals and pesticide residues. Another problem is biodiversity disruption and loss of the natural landscape, modified by the creation of new crops. Lack of attention to soil culture entails a number of unfavorable changes: chemical contamination, dehumification, soil fatigue, and, consequently, erosion [Rosati *et al.*, 2020].

Many of the practices used in traditional agriculture contribute to irreversible damage to the environment, which will significantly limit the possibility of food production in the future. This has led to an increase in crop acreage, and thus – the creation of many monocultures and the disruption of the natural biocenotic balance in agrarian ecosystems. In addition, as a result of intensive agricultural production methods, the threat of accumulation of organophosphate pesticide metabolites in plant and animal tissues, which are taken up with food by humans, producing irreversible health effects, is increasing [Giller *et al.*, 2017; Ponti *et al.*, 2012].

Globalization at the time of the greatest development of agriculture, led to the unification of agrarian systems. Over the course of several hundred years, food production has begun to focus on increasing food availability, which means higher yields and the elimination of crop-damaging factors. In order to achieve this, modern agricultural equipment and autonomous vehicles were introduced, which allowed for increased labor efficiency and accuracy of operations [Kwiatkowski & Harasim 2019; Weymann 2017]. Despite the advantages brought by the use of state-of-the-art technologies, frequent tillage operations can lead to serious damage to the soil or its individual layers (for example, the formation of a plow sole). The consequence can also be the entry of fertilizers into the deepest layers of the soil and the formation of permanent complexes between hazardous substances and naturally occurring substances in the soil, thereby increasing their uptake by the plant root system. These risks have shown the need for greater care about food quality in terms of the health safety of buyers, as well as the impact of agricultural production on the natural environment [Kwiatkowski & Harasim 2019; Seufert *et al.*, 2012].

According to Sanaullah *et al.* [2020] conventional agriculture systems are defined as being 'based on intensive use of agrochemicals to maximize agricultural productions', and encompassing 'intensive tillage to manipulate the soil physical properties and to control weeds, monocropping, and limited recycling of materials'. It was estimated that in 2015/2016, the agricultural sector depicted accounted for 87.5% of global cropland [Kassam *et al.*, 2019]. Reasons for the shift away from conventional agriculture to sustainable rural development in the 1990s included overproduction of food, environmental pollution caused by excessive chemicalization and mechanization of agriculture, and the dying of rural areas. In order to prevent these trends, under the Maastricht Treaty of 1992, the European Union recognized regulations promoting the production of environmentally related quality food. The following rules were implemented:

- agriculture is to provide food and at the same time provide environmental services;
- agricultural activities are to reduce environmental damage;

- agriculture is to strengthen farming structures that benefit the environment;
- agriculture is to strengthen the protection of natural resources, as well as biodiversity and landscape [Kılıç et al., 2020].

European Union legislation allows the combination of conventional and organic systems within a single farm, but regulates in detail the conditions that must be met in such a case. The solution, which is a mixed system, raises a lot of controversy – it questions, among other things, the legitimacy of collecting subsidies for organic production by the farmer. It is also considered in the aspect of action for the protection of the environment, biodiversity, as well as, importantly, the quality of the produced crops, which enter the market of organic products. Operating in both areas at the same time can also cause problems for inspection activities carried out by the respective certification body. In such a case, the separation of conventional and organic crops from each other and the storage of crops from both systems must be carefully checked [Nachtman 2015].

Farming in a conventional system does not have to mean "danger" or "harm" to crop quality and food security. As noted by Giller et al. [2017], considering conventional farming solely in terms of maximizing yields and associated farmer income is becoming obsolete. Conventional agriculture is evolving towards sustainable farming, focused on caring for soil health and crop quality. Thus, farming systems (conventional and organic) should not be confronted with each other in order to demonstrate the relative superiority of one of them. Instead, it seems expedient to develop increasingly effective management in each farming system (environmentally and economically sustainable) to contribute to the production of high-quality food, whether organic or conventional [Shennan et al., 2017].

Harasim & Kwiatkowski [2020] also present the view that conventional agriculture should evolve towards integrated production, that is, a way of farming that enables the realization of economic and ecological goals through the conscious use of modern production techniques, the systematic improvement of management and the implementation of various forms of biological progress in a manner conducive to the realization of these goals.

Environmental degradation caused by conventional management manifests itself primarily in the deterioration of soil, water and air quality (decline in soil fertility, pollution of ground and surface water mainly with nitrogen and phosphorus compounds, greenhouse gas emissions into the atmosphere, mass extinction of many plant and animal species, contamination of agricultural crops with residues of the chemicals used). These threats have shown that more attention should be paid to food quality in the context of consumer health safety, as well as to the impact of agricultural production on the environment, so as not to lead to the degradation and irreversible destruction of the natural basis of food production in the future [Kılıç et al., 2020].

The afore-mentioned progressive degradation of the natural environment, in terms of the use of natural resources in food production, makes it necessary to consider environmental aspects throughout the agri-food chain. The food sector, after the energy sector, is a major determinant of global sustainability. Its efficient and effective operation should minimize the use and degradation of the natural environment and the loss of biodiversity [BCFN 2014a].

The International Organization for Biological and Integrated Control of Harmful Animals and Plants (IOBC) has defined integrated production as: an agricultural system for food production that optimizes the use of natural resources and regulatory mechanisms, ensuring sustainable agriculture in the long term. The system involves the careful selection of biological methods, cultivation techniques and chemical processes with the goal of balancing environmental, profitability and social requirements. Essentially, it is a voluntary model based on the practical and continuous application (through the exchange of knowledge and experience between technical services, farmers and farms) of technological innovations and tools that, if used effectively, make it possible to achieve the quality, safety and environmental standards demanded by today's society [European Economic and Social Committee 2014; Hoang 2021].

The concept of integrated production is often used as a synonym for integrated agriculture. Although these are parallel production systems with many common elements, they are different in nature and represent diverse models from which the farmer can choose. Integrated production is based on a sectoral vision that includes different provisions depending on the product, while integrated agriculture refers to the entire management of the farm. Integrated production includes ecological, ethical, social aspects of agricultural production, as well as the issue of food quality and safety. In fact, integrated production is considered one of the most relevant international standards for food production [European Economic and Social Committee 2014].

The production of high-quality fruits and vegetables that makes sustainable use of technical and biological advances in cultivation, plant protection and fertilization, minimizing the undesirable side effects of the agrochemicals used, with particular attention to protecting the environment and human health is called integrated plant production (IP). This type of production constitutes a modern food quality system. In order to promote and protect consistent principles of integrated production in the European Union, the European Initiative for Sustainable Development in Agriculture (EISA) was established in 2001. One of EISA's first tasks was to create the European Code for Integrated Production, which served as the FAO's basis for defining sustainable practices in agriculture. The products produced under an integrated production system can obtain a guarantee label, but only if they meet general standards and technical standards specific to each crop [EC 2020a].

The continuous increase in agricultural productivity by increasing the use of agrochemicals is causing damage to the natural environment [Gulbicka & Kwasek 2006]. Global agribusiness concerns are concentrating agricultural production, finding a market for fertilizers, pesticides, feed, veterinary drugs, machinery, etc. Industrially produced food can contain, despite controls, many chemical contaminants: nitrates and nitrites, heavy metals, pesticide and herbicide residues, substances that delay or accelerate ripening, drug residues, substances that reduce spoilage during storage and transport, hormones, mycotoxins, etc. [EC 2020b].

Considering the development of agriculture throughout history and currently, it can be noted that in the 21st century there has been a successive return to tradition – through the protection of biodiversity or a return to old plant varieties and animal species. On the other hand, transgenic crops (GMOs) – formally cultivated since 1994 (soybeans, corn, cotton, canola), mainly in the US, Argentina, Canada and China – are being promoted. However, since the 1990s, methods of integrated crop production, combining biological and chemical methods of plant protection, have been implemented into agricultural practice. The idea of sustainable development was formulated in 1987 in the report "Our Common Future," produced by the UN Commission on Environmental Affairs and Development, which was headed by Gro Harlem Brundtland. "Human beings have the right to use nature's resources, but they also have a duty to protect them so as not to compromise the chances of meeting the needs of future generations" [Schmit et al., 2020].

This is a great achievement of modern agriculture, and it should be forecast that an increasing number of number of farms will implement the principles of the so-called Good

Agricultural Practice (GAP). Good Agricultural Practice is a set of practices that allow producing safe food using all available methods and means. Under EU recommendations, it has been made mandatory for farmers applying for financial support. GAP is a set of rules of conduct, the application of which gives the farmer confidence that his work, while producing the desired result, does not simultaneously cause damage to nature, or that the negative impact of agriculture on the environment is as small as possible.

GAP involves the farmer's compliance with applicable environmental laws and regulations environmental laws, among others, in the field of:

— use and storage of organic and mineral fertilizers
— environmentally safe agricultural use of wastewater on the farm
— environmentally safe agricultural use of municipal sewage sludge
— proper use of plant protection products
— grassland management
— maintaining cleanliness and order on the farm
— protecting natural habitats
— protecting soils from erosion
— protecting water resources against pollution of agricultural origin [EC 2020a, 2020b].

Implementation of sustainable development principles is becoming a necessity. Intensive agriculture only achieves economic goals, while generating adverse externalities that threaten the environment. The continuation of this type of development path of the agricultural sector will prevent future generations from enjoying the full benefits of natural resources. This does not mean that the concept of sustainable agriculture is being implemented in countries or regions with extensive agriculture. This is because this paradigm implies a balance between three goals – economic, environmental and social. Thus, only a proper balance of these three aspects of farming can indicate the right direction for the development of any economic department, including agriculture. The strategy for protecting the environment in rural areas assumes that the farmer's application of the principles of good farming will effectively reduce the environmental risks generated by agriculture. In addition, many EU and national regulations apply to the farm. Adherence to the principles of GAP is closely related to the production by farms of high-quality agricultural crops that are the basis for healthy and safe food [Grossman 2011; Schmit et al., 2020].

The development of agriculture in the world is not only a pro-environmental (sustainable) trend, but also includes using the achievements of civilization in order to maximize profits from production. The development of so-called precision agriculture (the use of GPS technology in agriculture), the development of genetic engineering and plant breeding (GMOs, such as high-yielding varieties of wheat), the increase in animal productivity, the increase in the use of mineral fertilizers and plant protection products, the development of agricultural machinery, the development of soil science should be mentioned here [Gawęcki 2010; Lotz et al., 2018].

Sustainable agriculture is a basic need of civil society, which can be met through various production models, including organic farming and integrated farming referred to as alternative agriculture. Alternative agriculture, compared to conventional agriculture, does not seek high yields at the expense of heavy investment in agricultural chemistry. The goal of alternative agriculture is to produce food in a way that does not degrade the natural environment. This is important from both an economic (demand for organically produced food products) and an ecological point of view [Kharel et al., 2022].

Chapter 3

Chemisation of agriculture in relation to healthy and safe food

3.1 INTRODUCTION

Over the last decade, the approach to food security has changed many times. This discussion was initiated by the FAO's World Food Conference organised in 1974, during which "food security" was defined as "the availability at all times of adequate world food supplies, of basic foodstuff to sustain a steady expansion of food consumption and to offset fluctuation in production and prices" [Borch & Kjærnes 2016].

Despite that it provides nutrients necessary for the organism, food should be characterised by proper health quality. The health quality of food results from many factors that occur in the natural human environment, starting from conditions for obtaining raw materials and processing them. Quality is an important term in food production and assessment. The basic definition of quality is as follows: "the sum of the characteristics of a unit which influence its capacities in satisfying determined and expected needs" [Jackson & Fahrig 2015].

Food safety is of essential importance in public health protection and it is also one of the pillars of European food policy. Care for the safety of the final product should already start from the production of a raw material, animal husbandry conditions, and animal feed quality through production technology and retail networks [Kahl *et al.*, 2010].

Food safety and food quality are important factors that affect public health. The lack of food safety has a negative impact on the condition of each national economy and also on households by closing export markets or reducing the availability of food in the domestic market. Consumers have the right to safe food that will not cause a threat to their lives or health. Consumers' growing knowledge and awareness with regard to consumption of food and its linkages to health cause a greater demand for food products of high quality which is a result of specific methods of their production, their unique composition, and also their specific origin [Leventon & Laudan 2017; Moragues-Faus 2017].

Health food safety is defined by the Codex Alimentarius as "Assurance that food will not cause adverse health effects to the consumer when it is prepared and/or eaten according to its intended use" [Burchi & de Muro 2016]. According to the FAO [2021], food safety is "all conditions that must be met with respect to, in particular: additives and flavourings used, levels of pollutants, pesticide residues, food irradiation conditions, organoleptic characteristics and measures that must be taken at all stages of food production or marketing to ensure human health and life". From the consumer's point of view, health safety is a major quality characteristic and therefore European and global food law precisely regulates this issue [Burchi & de Muro 2016].

DOI: 10.1201/9781003380771-3

3.2 PESTICIDES USE IN AGRICULTURE

Persistent organic pollutants are a major group of environmental pollutants. Pesticides, among others, belong to this group. Due to the frequency of their use as well as their environmental persistence, toxicity and ability to bioaccumulate, they are one of the most toxic substances that contaminate the environment. Their presence is dangerous mainly in fruits and vegetables since they are one of disease risk factors for human health. However, they protect agricultural crops against damage done by pests. As reported by Roy [2002], the losses of crops caused by pests and plant diseases are quite high both in developed and developing countries. These are reported to be in the range of 10–30% in the former case and 40–75% in the latter one.

Pesticides are included in different groups of chemical compounds and the frequency of their use is associated with the wide range of their effects [Aktar *et al.*, 2009; FAO 2021; Neef *et al.*, 2012]. Pesticides are artificially synthesised chemical substances or their mixtures. They are popularly called "plant protection products" [Rani *et al.*, 2021].

According to the International Code of Conduct on Pesticide Management, "pesticide means any substance, or mixture of substances, intended for repelling, destroying or controlling any pest, including vectors of human or animal disease, unwanted species of plants or animals causing harm during or otherwise interfering with the production, processing, storage, transport, or marketing of food, agricultural commodities, wood and wood products or animal feeding stuffs, or which may be administered to animals for the control of insects, arachnids or other pests in or on their bodies. The term includes substances intended for use as insect or plant growth regulators; defoliants; desiccants; agents for setting, thinning or preventing the premature fall of fruit; and substances applied to crops either before or after harvest to protect the commodity from deterioration during storage and transport" [Bhadekar 2011; Chopra 2011; Rani *et al.*, 2021; Sassolas 2012].

The basic measure that ensures that the negative impact of pesticides on health is reduced is to provide continuous control of their presence and also to determine the highest permitted limits for residues of these compounds in different environmental segments. Improper use of pesticides by farmers as well as the use of their excessive amounts or failure to observe pre-harvest intervals have a decisive effect on pesticide residues in plants [Amvrazi 2011; Grewal *et al.*, 2017; Ssemugabo *et al.*, 2022]. In agriculture, pesticide use increases with the growing global population as well as with industrial progress and scientific and technical progress. This is a result of continuous search for methods to increase the productivity of crops and to reduce crop losses by, for example, excluding plant diseases or reducing populations of insects that feed on crops. In spite of many advantages arising from the use of pesticides, the main problem is their harmful effects on human and animal health [Bajwa and Sandhu 2014; Chung 2018; European Commission 2020].

Pesticides are classified based on active ingredients, chemical structure, mode of action and toxicity [Botitsi *et al.*, 2017]. There are organic and inorganic pesticides. Organic pesticides are carbon based, either natural pesticides from natural occurring materials or synthetic pesticides, synthetically produced from organic chemicals. Inorganic pesticides are derived from mineral compounds that occur as deposit in nature, mainly compounds of antimony, copper, boron, fluorine, mercury, selenium, thallium, zinc, phosphorus and sulphur [Patinha *et al.*, 2018; Sarwar 2016].

Depending on the character of their use, the following can be distinguished, among others: fungicides, herbicides and insecticides. The most important and most frequently used pesticides are fungicides. They are used in agriculture to prevent or control crop diseases caused by fungi. There are many methods for classification of fungicides, among others based on the targeted place of action of the compound in the pathogen cell, the chemical structure of fungicides and also their toxicity. Fungicides are also classified in terms of the chemical structure of the compound being the active ingredient of the pesticide product applied. The most important groups used in agriculture and fruticulture include, among others, triazoles, which are the most numerous group of fungicides, strobilurin, dithiocarbamates, phthalic acid imides and benzimidazole compounds [Nicolopoulou-Stamati et al., 2016]. Fungicides are also divided based on their degree of toxicity. The class of entirely toxic compounds includes: captafol, hexachlorobenzene, mercuric chloride and phenylmercury acetate, which are compounds with antifungal properties [The WHO 2010].

Herbicides are chemical compounds that have the ability to destroy or inhibit the growth and development of plants. Weed-killing agents find application in horticulture, forestry, land drainage activities and plant control activities in industrial areas as well as on lawns, pavements and roadsides. The greatest demand for herbicides can be found in agriculture and it is now difficult to imagine farming without herbicide use. Herbicides penetrate into plants through their leaves and/or roots and subsequently are distributed through the system of vascular bundles, causing disruptions in life processes. Their activity is sometimes entirely local and restricted to the place that the spray solution ultimately reaches. The great majority of herbicides have selective effects, affecting only a specific group of plants, i.e. particular dicotyledonous or monocotyledonous species. Herbicide selectivity provides the possibility of carrying out spraying in plantations during the growing season without destroying crop plants. There are also non-selective herbicides that are capable of doing damage to practically every type of vegetation. Herbicides are applied in fields to reduce crop losses caused by competitive effects of weeds. This is of important economic significance and that is why the use of these chemicals has become an important element of farming. Nevertheless, herbicide treatments can be a source of pollution of plant products and soils, among others, and therefore the level of herbicide residues should be monitored both in the soil environment and in agricultural commodities. Exceedance of the permissible herbicide residue limits in plant products may cause a negative impact on consumers' health [Deadman 2017; Hedlund et al., 2019; Patinha et al., 2018; WHO 2010].

Insecticides include inorganic chemical substances, compounds of plant origin and synthetic compounds of organic origin. A huge amount of these compounds belong to strong poisons. Insecticides are divided depending on their chemical structure, application mode and place, power to destroy pests and application method. They can be applied by spraying, glue trapping, fogging, dusting, broadcasting and dressing. Therefore, they are inside a plant or on its surface. An insect can be affected by an insecticide through its exoskeleton (externally) or through ingestion or inhalation. According to their mode of distribution on the plant, insecticides are divided into compounds that are systemic insecticides and surface (contact) insecticides. Substances that act systemically are distributed in the plant thanks to the vascular system regardless of the place where the substance penetrated into the plant. It is distributed to all parts of the plant, whereas surface insecticides protect the plant's outer parts against pests. Insecticides have a toxic effect on insects and therefore there is also a risk of toxicity to mammals [Eurostat 2020a; Lee et al., 2019; Jacqued et al., 2022; Sarwar 2016].

Different food commodities have varying compositions and properties based on their nature and classification, which is also applicable to pesticides. Various factors influence the level of pesticide absorption, penetration, and degradation, and these factors vary across food properties. The movement and dissipation rate of pesticides is closely linked to the physiochemical traits of the pesticide and the surrounding environmental conditions [Bajwa & Sandhu 2014]. The measure of pesticide resistance to degradation under different conditions is its "half-life" value. This parameter refers to the time it takes for half amount of the pesticide to break down, and ranging from a few hours or days to several years for highly persistent ones [Helfrich 2009].

Additionally, pesticides are classified based on the acute oral and dermal toxicity to the rat as the standard toxicology procedure, presented as LD50 or LC50 values. Given in milligrams of pesticide per kilogram of body weight or in parts per million (ppm), LD50 and LC50 values are determined based on a single dosage and help compare the toxicity levels of different formulations and active ingredients. The products having lower values of these parameters considered more toxic to humans and animals. When used as directed, pesticides with high LD50 values are least harmful to humans [Hock 2022].

According to USA regulations, there are four Toxicity Categories for acute hazards of pesticide products [Code of Federal Regulation, Title 40 2023]. A Toxicity Category is assigned for each of five types of acute exposure, as given in the Table 3.1.

Table 3.1 Acute toxicity categories for pesticide products [Code of Federal Regulation Part 156 2023].

Routes of Exposure	Toxicity Cat. I	Toxicity Cat. II	Toxicity Cat. III	Toxicity Cat. IV
Oral LD50 [mg/kg]	≤ 50	50–500	500–5,000	>5,000
Inhalation LC50 [mg/dm^3]	≤ 0.2	0.2–2	2–20	>20
Dermal LD50 [mg/kg]	≤ 200	200–2,000	2,000–20,000	>20,000
Eye effects	Corrosive corneal opacity not reversible within 7 d	Corneal opacity reversible within 7d; irritation persisting for 7 d	No corneal opacity; irritation reversible within 7 d	No irritation
Skin effects	Corrosive	Severe irritation at 72 h	Moderate irritation at 72 h	Mild or slight irritation at 72 h

Different category of toxicity requires a different labeling of the product. Pesticides that are highly toxic (Toxicity Category I) must have the signal words "DANGER", additionality, if the product is categorized as Toxicity Category I due to its oral, inhalation, or dermal toxicity (not including skin and eye irritation), it should display the word "Poison" in red on a background of significantly contrasting color. The oral LD50 for these pesticides ranges from a trace amount to 50 mg/kg, which means even a small amount could be fatal if ingested. Some pesticides labeled as DANGER may not have the skull and crossbones symbol because they can cause severe skin and eye effects. Moderately toxic pesticides (Toxicity Category II) are labeled with the signal word "WARNING". The oral LD50 for this category ranges from 50 to 500 mg/kg, so even a small amount could be fatal if swallowed. Pesticides classified as slightly toxic or relatively nontoxic (Toxicity Categories III and IV) are labeled with the signal word "CAUTION".

The chronic toxicity of a pesticide is evaluated by subjecting test animals to the active ingredient for an extended duration. Adverse effects that occur as a result of repeated small doses are known as chronic effects. These effects can include birth defects, tumor development, blood disorders, and nerve disorders. Determining the chronic toxicity of a pesticide is more challenging with laboratory analysis compared to acute toxicity [Hock 2022].

Although most of the pesticides classification is based on the acute oral LD50 value, dermal toxicity is always considered since pesticide handling takes a high proportion of dermal exposure [Strickland *et al.*, 2018]. Moreover, according to Giusti *et al.* [2018], pesticides concentration in the air exposes community through inhalation. For instance, the use of the median lethal dose (LC50) which defines the average concentration of chemicals as gas, vapour, mist, fume or dust is capable of killing 50% of the test animals exposed by inhalation under specific experimental conditions, expressed as milligram (mg) per litre (l) over a given period of exposure [Deadman 2017]. Pesticide formulation may contain more than one ingredient such as wetting agent of significant toxicity and then the classification should correspond to the toxicity of mixed ingredients [World Health Organization 2010]. Based on the active and inert ingredients and target pests, pesticides have different modes of action: interfering with amino acid and protein synthesis, nervous system, cell division, energy production, respiration, growth or development regulation, photosynthesis, deoxyribonucleic acid (DNA) damage and methylation, membrane integrity or multisite, and sometimes with unknown specificity [Maul *et al.*, 2018].

After pesticides are sprayed, they gradually begin to dissipate in the environment. Each pesticide used on crops requires a specific waiting period before harvesting, which can vary between different pesticides and crops. Only after this waiting period has passed are the food products considered safe for consumption. Harvesting fruits and vegetables before the waiting period has ended can result in higher levels of potentially harmful residues (Bajwa & Sandhu 2014). According to Gupta [2006] despite the restrictions and regulations on pesticide use, India accounts for one-third of pesticide poisoning cases in the world. Excessive pesticide residues can lead to various harmful consequences such as blindness, cancer, liver and nervous system diseases, and more. Over time, these effects can worsen and expand to causing infertility, elevated cholesterol levels, increased infant mortality rates, as well as various metabolic and genetic disorders [Gupta 2006].

Bajwa & Sandhu [2014] reported that pesticide residues, left to in the raw materials after harvesting, are beyond the control of the consumer but have a deleterious effect on human health. The crops treated with pesticides may contain an unpredictable amount of pesticides, and therefore it is important to find the methods for removing these chemicals from food.

The presence of pesticides in food is influenced by the specific characteristics of the pesticide, the type and portion of food, and environmental factors.

Washing with water or soaking in salt or chemical solutions, such as chlorine or hydrogen peroxide, have been proven to greatly reduce pesticide levels in the food products. Additionally, peeling fruits and vegetables can remove pesticide residues from the outer layers. Thermal treatments like boiling, steaming, pasteurization, blanching, and canning can effectively degrade pesticides. However, preservation techniques like drying or dehydration can actually increase pesticide content in food due to the concentration effect [Bajwa & Sandhu 2014; Iizuka & Shimizu 2014; Zhang *et al.*, 2022].

Many other techniques like refining, fermentation and curing have been reported to affect the pesticide level in foods to varied extent. Milling, baking, wine making, malting and brewing resulted in lowering of the pesticide residue level in the end products [Bajwa & Sandhu 2014; Regueiro *et al.*, 2013]. In addition, cold storage and post-harvest treatments have been proven to be effective. However, the extent of reduction depends on the initial pesticide concentration at the time of harvest, the raw material or food product composition being treated, and the specific type of pesticide.

Substances used as plant protection agents as well as their metabolites or degradation products can be pesticide residues remaining in food. Pesticides can also penetrate into food through direct application of plant protection products on vegetables, whereas in products of animal origin as a result of ingestion of contaminated feed by animals, in particular. Protection of food during storage and transport is related to the occurrence of pesticide residues. Decrease in organoleptic characteristics and nutritional values is caused by the presence of pesticides in food. It has also been proven that their content is reduced with storage time and during processing. Pretreatment operations and technological processing of raw materials result in reduction of pesticide residues [Tomer & Sangha 2013; Chavarri *et al.*, 2005].

The location of pesticides in different parts of food differs depending on the type of the molecule, type of the food product and environmental conditions. A pesticide can be degraded due to photolysis, hydrolysis, oxidation and reduction, metabolism, temperature, and pH. The level of pesticide residues is affected by washing, preparatory steps, heating or cooking, processing during the manufacture of a product as well as by post-harvest handling and storage. The extent of reduction varies depending on the type of the pesticide molecule, place of location, type of commodity, processing stages and the product being prepared. Washing of raw materials is the simplest way to reduce the pesticide residue in the end product. A more effective and convenient alternative can be washing with chlorine water or with diluted solutions of other chemicals depending on the food product [Bajwa & Sandhu 2014]. The authors drew this conclusion based on an extensive literature review, from which some examples are reported.

According to Wu *et al.* [2007], treatment of four pesticides: cypermethrin, diazinon, parathion, and methyl-parathion included in vegetables, with low-concentration ozonated water with initial content of 1.4 ppm was able to remove 60–99% of the initial concentration of the pesticides within 30 min. The removal of cypermethrin was the most effective. The efficiency of pesticides degradation was highly depended on the dissolved ozone concentration and temperature. Higher temperature enhanced the efficiency of pesticides removal. For example the maximal efficacy for diazinon removal was found at temperature 15–20°C.

Kim *et al.* [2000] treated soybeans with 0.3 ppm ozone water for 30 min. The study analyzed the impact of different soaking/ozonation techniques on the residues of carbendazim, captan, diazinon, fenthim, dichlorvos and chlorpyriphos. Results showed that ozone treatment was more effective in eliminating pesticides compared to soaking in water, with captan showing the highest susceptibility to degradation and chlorpyriphos showing the least. According to Zhanggui *et al.* [2003], wheat and corn biomass treated with 15–20 ppm ozone experienced reduced pesticide residue concentration, with a maximum degradation rate of 12.3% per day.

According the study of Cabras & Angioni [2000] which examined the pesticide concentration in grapes and their processing products, the concentrations of the residues of such pesticides as benalaxyl, metalaxyl, phosalone, and procymidone in sun-dried grapes was similar to those observed in the fresh grapes. The differences were found in the cases of iprodione, which concentration was 1.6 times higher, as well as vinclozolin and dimethoate, which concentration were approx. 30% and 20% lower, respectively. They observed that oven-drying process did not influence the content of benalaxyl, metalaxyl, and vinclozolin both in the fresh and dried fruit while it lowered the residue levels of iprodione and procymidone in raisins compared to the fresh fruit.

The residues of herbicides, insecticides and fungicides during processing of barley into malt, as studied by Navarro *et al.* [2007], showed that pesticide concentration declined along the process, although in different proportions. The carryover of residues after steeping was 45–85%. Significant correlation ($r > 0.92$) was observed between percentages removed after steeping and the POW values of pesticides. The amount remaining after malting ranged from 13 to 51% for fenitrothion and nuarimol, respectively. Steeping was found to be the most important stage in the removal of pesticide residues (52%), followed by germination (25%) and kilning (drying and curing, 23%). During malt storage (3 months), the fall in pesticide residues was not significant. Applying the standard 1st-order kinetics equation (r > 0.95), the half-lives obtained for pesticides during malt storage varied from 244 to 1,533 days for myclobutanil and nuarimol, respectively. In beer, the production process leads to a reduction of residues present in grain and hops used in brewing.

Research by Chavarri et al. [2005] and Lennon et al. [2006] conducted on various food products showed that a combination of washing and blanching led to >50% reduction in pesticide residue levels in all samples except peaches. Total amounts of pesticides removed by all of the combined canning operations ranged from 90 to 100% in most products. Pepper retained 61% of chlorpyriphos, but these residues disappeared during storage of cans for 3 months. Acephate showed a surprising tenacity in peaches, as 11% of the original residues were still present in cans stored for 2 years. There was a significant difference (P = 0.05) in residue concentrations of the parent compounds between the whole peach, the half peach with peel, and the half peach without the peel, during the commercial canning process. However, there were no significant differences in residue concentrations found between the half peach without the peel and the canned peach samples. The dissipation rate of metabolite residues varied from compound to compound.

Ruiz-Mendez et al. [2005] investigated the effects of various processing stages of olive oil refining on pesticide levels. Oil samples were spiked with endosulfan (α-endosulfan, β-endosulfan and endosulfan sulfate), simazine, oxifluorfen and diflufenican and subjected to physical refining (bleaching and deodorization). Bleaching using earths was effective only for the elimination of simazine. For removal of the other pesticides tested, a physical refining treatment was required at 240°C in the deodorizing stage for a period of 1 to 3 h. Fukazawa et al. [2007] studied the behaviour of N-methylcarbamate pesticides during refinement processing of spiked (5 ppm) soybean oil. By degumming, aldicarb, aldicarb sulfoxide, aldicarb sulfone, oxamyl, thiodicarb, carbosulfan and benfuracarb contents decreased by up to 70% with H_3PO_4 treatment and decreases of <26% were noted for the other pesticides. With hot water treatment, the decrease in the content of any of the pesticides was <52%. With alkali refining, the decrease varied with the pesticide in the range 8–100%. NaOH treatment was found effective in removing pesticides. With bleaching, aldicarb, aldicarb sulfoxide, aldicarb sulfone, oxamyl, methomyl, thiodicarb, carbosulfan, benfuracarb, bendiocarb and furathiocarb contents were decreased by >80%, with activated clay containing activated charcoal. Carbaryl content was decreased markedly using this clay. With deodorization, 40% furathiocarb, 14% carbosulfan, 11% benfuracarb and 3% carbofuran could still be detected following deodorization at 260°C. Degumming with H_3PO_4 and bleaching with activated clay caused conversion of carbosulfan and benfuracarb to carbofuran.

Tomato and carrot pulp contained a higher percentage of all pesticide residues, except for mancozeb in tomato juice. Although there was a difference in the relative distribution of the pesticides between the commodities with greater amounts present in the pulp of tomatoes, the pesticides followed a similar trend in both. Pesticides with the highest water solubility were present to a greater extent in the juice. An exception was noted in the case of diazinon and parathion, which were present in higher amounts in the pulp than their water solubility suggested. The residue in the pulp ranged from 56.4 to 75.2% for carrots and 49.7 to 95.4% for tomatoes. Washing of the produce removed more residues from carrots than from tomatoes, but it did not affect the relative distribution of the residues [Burchat *et al.*, 1998]. The behaviour and fate of the chemical varied with the pesticide as well as the crop. Pesticide residues were greatly decreased in tomato juice under cold or hot break. A sharp decline in profenofos level was noted after treatment by pectinex ultra SP-L and benzyme M during tomato crushing [Romeh *et al.*, 2009].

Special precautions should be taken to remove pesticides from raw materials intended to be used for preparation of concentrated and dehydrated products. There is an urgent need to monitor pesticide residues in order to standardise application doses. Equally important is to develop or find new pesticide molecules with high effectiveness and fast degrading capabilities. Pest management is one of the major inputs in agricultural production; therefore, this area needs great attention in order to economise the production, to provide safe foods, and to lower medical expenses for treatment of resulting ailments [Bajwa & Sandhu 2014].

It is impossible to completely eliminate artificial chemical compounds found in foods supplied to consumers and therefore permissible doses accepted by the organism have been established for individual pesticides. An acceptable dose is the maximum pesticide residue which remains in raw materials and products after the pre-harvest interval. In spite of that, producers and processors of agricultural commodities, seeking to increase yields and thus their income, frequently do not observe the applicable limits [Anastassides *et al.*, 2002; Keikotlhaile *et al.*, 2010].

Today, pesticides have become the cornerstone of the predominant agricultural systems. In the European Union (EU), pesticide sales reached 370 million kilograms in 2018 [Eurostat 2020a]. The pesticides sold the most (by mass) are fungicides (46%), followed by herbicides (35%) and insecticides (11%) [Eurostat 2020b]. Pesticides and other technological advances of the Green Revolution have enabled farmers to drastically increase crop yields and countries to improve food security [Hedlund *et al.*, 2019]. However, reducing pesticide use has become a goal shared by several countries and a major issue in public policies [Lee *et al.*, 2019] since negative impacts of pesticides on the environment and on human health have been demonstrated unambiguously.

Although pesticide residues below MRLs in food are considered non-hazardous to human, more research on chronic exposure is suggested because most of the pesticides are fat soluble and accumulate in body tissue over time. Continuous investments in research are necessary to develop new non-persistent active chemicals for effective pest control. It is crucial to regularly educate consumers and all participants in the food production industry about the proper and safe usage of pesticides.

This should involve providing easy access to updated information regarding pesticides. Communities, especially those employed in pesticide factories and agriculture industries, should be cautioned about the ways of being exposed to pesticides, i.e. contact with the skin and eyes or inhalation. The international, regional, and national authorities have a duty to ensure that food is safe and free from pesticide residues, and to establish systems for tracking both raw and processed food. Furthermore, the consumers themselves are responsible for ensuring the safety of their food by obtaining it from a trustworthy source and following safe food preparation guidelines [Yamada 2017; Zikankuba *et al.*, 2019].

3.3 FERTILISERS AS AN AGRICULTURAL CHEMISATION FACTOR

Fertilisers are products intended to supply nutrients to plants or to increase soil fertility. We distinguish mineral and natural (organic) fertilisers, among others. Fertilisation is an important yield-enhancing factor and also one of the main indicators for agricultural production intensity and effectiveness. Mineral fertilisers used in the agricultural production process have a great impact on plant growth and quality and enrich the soil with necessary minerals. Fertilisation has a huge effect on yield quantity. The following mineral fertilisers can be distinguished: nitrogen, magnesium, calcium, potassium and phosphorus fertilisers. Nitrogen fertilisers influence the development of plants and contribute to their intense growth, increasing their biomass and seed yield. Malnutrition of crops with this element leads to disruptions in their functioning, causes many adverse physiological effects and in consequence results in reduced yield [Franco & Delgado 2022; Katan 2009]. Magnesium fertilisers have an important plant nutrient, i.e. magnesium, which is necessary to build, e.g., chlorophyll. Magnesium deficiency in plants causes their negative growth and low yield. This nutrient participates in the photosynthesis process and is an activator of a number of enzymatic and biochemical processes. Magnesium reduces plant susceptibility to diseases and also stimulates the yield-forming effect of nitrogen. Calcium fertilisers belong to artificial fertilisers from which plant-available ions are released and which reduce acidic pH of soils. Potassium fertilisers contain potassium in the form of soluble salts that are available to plants. Phosphorus fertilisers have an enormous impact on yield quality, while to a smaller degree on yield quantity. This mineral positively affects tillering of cereals and rooting of plants [Aktar et al., 2019].

Natural fertilisers perform an important production and ecological role because they contain all nutrients necessary for plants and organic matter, which is a substrate for humus formation in soils. Fertilisers that are greatly valued in farming are as follows: farmyard manure, liquid manure and poultry manure. The amount of nutrients in these fertilisers is different and depends on the following, among others: livestock breed, nutrition system, intended use of livestock, livestock breeding method, quality and quantity of bedding used, and also manure storage method. Farmyard manure is a mixture of animal faeces and urine with bedding material. This fertiliser has versatile effects since it contains all nutrients necessary for plants. Farmyard manure may significantly reduce the total cost of fertilisers used since it is the cheapest among all known fertilisers, but it is not sufficiently active to replace mineral fertilisers. Liquid manure differs from solid manure not only in its physical properties but also in its fertilising effects and chemical composition. It is a more aggressive fertiliser, it has a faster effect on the soil, and also, unlike solid manure, it is liquid. Moreover, its utilisation by plants is very fast because most fertilising substances occur in mineral form, e.g. nitrogen from liquid manure is better absorbed by plants than that from solid manure. Liquid manure is a mixture of urine, faeces and water runoff from washing stalls. It is produced in stalls adapted to keeping livestock without bedding. Organic fertilisers are produced from organic raw materials of plant origin (straw, leaves, grass, sawdust) or of animal origin (urine, faeces), or from a mixture of organic matter, including composts [Kakar et al., 2019; Zaman et al., 2018].

According to the findings of Wan et al. [2021], the use of chemical fertilisers for high citrus yields is globaly commonly used, particulary in China, leading to soil deterioration in citrus orchards. To address this problem, the authors suggest that organic fertilisers offer a promising solution. Based on the results of the study they concluded that a combination of organic and chemical fertilisers, as well as bioorganic and chemical fertilisers, could improve soil fertility, enhance citrus growth physiology, reduce NO_3^--N leaching, and increase yields.

As reported by Krasilnikov et al. [2022] according to Luo et al. [2021], biobased nitrogen fertiliser derived from animal manure can substitute synthetic mineral N fertilizer and contribute to more sustainable methods of agriculture. Orden et al. [2021] have found that organic solid wastes derived from onion production can be successfully used as soil fertilizer.

They noticed that this sector of agriculture in the low valley of Río Colorado (Buenos Aires) generates 12,000–20,000 Mg/year of vegetal wastes, and causes many problems with these waste management. According to the results, the compost applied influenced soil pH, electrical conductivity, and nutrient levels. The soil enzymes were already at high levels before the compost was added, which could explain the minimal effects on microbial soil activity.

Oueriemmi et al. [2021] found in their study that the availability and cost of farmyard manure, a traditional amendment for enhancing sandy soil fertility in dry climates, is diminishing. To analyze the impact of three different organic residues on barley yield, nutrient uptake, and certain soil properties, a field trial was conducted with a single application of these residues. The effects were evaluated after two consecutive harvests. The authors stated that adding the municipal solid waste compost, sewage sludge compost or farmyard manure increased content of organic matter, cation exchange capacity and available nutrients (P, K, Mg, Ca) in the soil, and the barley grain yield.

Several more studies have documented the potential advantages of organic fertilizers. In these studies, researchers observed an increase in soil microbial activity, resulting in improved crop growth and the suppression of pests and diseases [Chang et al., 2010; Lin et al., 2019; Zhang et al., 2012]. The researchers have found that tea grown with bio-organic fertilizers exhibits better color and taste in comparison to tea that has been treated with chemical fertilizers [Lin et al., 2010; Zhang et al., 2012]. Studies have also suggested that the use of organic fertilisers resulted in higher seedling biomass and significantly improved the soil fungal to bacterial ratio as well as soil enzyme activity [Sun et al., 2017]. Application of organic fertilizer has also other beneficial effects. Prolonged usage of chemical fertilisers can result in soil acidification, degradation of the micro-ecosystems in the rhizosphere and nutritional imbalance. Additionally, it amplifies the presence of heavy metal ions in the soil. The application of organic fertilisers can mitigate these unfavourable changes, and consequently enhance plant productivity [Li et al., 2018].

3.4 SUSTAINABLE (INTEGRATED) CHEMISATION OF AGRICULTURE

Sustainable development primarily involves meeting current social needs and future generations' needs, while preserving the natural environment in the best possible condition. Agriculture should seek to be, to a large extent, sustainable (stable), ecological (it should not cause pollution) and socially oriented (it should guarantee decent living conditions for the population). Sustainable farming provides food produced using a minimal amount of fertilisers and crop protection products. It is oriented towards using the earth's resources in a way that does not destroy their natural sources and despite of that enables basic needs of successive generations of producers and consumers to be met [Giusti et al., 2018; Kwasek & Obiedzińska 2014].

In choosing foods, attention is paid more and more frequently to the way in which they were produced. Consumers expect fresh vegetables and fruits, free from pollutants and pesticides. Unfortunately, excessive use of plant protection chemicals to easily control agricultural pests and protect yields often causes a threat to the natural environment. Therefore, the possibility increases that their residues will be left in products of plant origin in amounts harmful to health. This situation induces us to seek new production technologies and methods that will allow economically profitable yields to be obtained, produced in a way that minimises the risk to the environment. A system that meets this requirements is integrated plant protection, which allows safe and healthy food with high quality parameters to be produced and which also protects the natural environment. The basis for this system is to use sustainable technical and biological progress in crop cultivation, protection, and fertilisation [Glavic *et al.*, 2021]. Integrated plant protection uses to the greatest possible extent natural physiological processes of the plant supported by rational use of conventional, natural and biological plant protection products. The following three steps are used in the integrated plant protection system: prevention, monitoring and control. Prevention of the occurrence of harmful organisms or reduction of their negative effects should be achieved by using appropriate crop rotation, proper agronomic practices, selection of resistant or tolerant cultivars, and the use of sustainable fertilisation, irrigation, and liming. Reducing the possibility for agricultural pests to colonise crops, creating conditions promoting the occurrence of beneficial organisms and observing phytosanitary hygiene principles play an important role. Monitoring the development of diseases and pests in crops must be the basis for making decisions concerning the type of treatment. Biological methods are used to control harmful organisms, while the use of plant protection chemicals is reduced to the necessary minimum [Barzman *et al.*, 2015; Marrone 2019].

An important element of a sustainable pesticide use strategy is fertilisation that provides continuous soil fertility by creating appropriate living conditions for soil microorganisms. Soil that is well aerated and has proper pH and an adequate amount of organic substances provides favourable living conditions for plants that are resistant to variable climatic conditions as well as to diseases and pests. Sustainable fertilisation has also a huge impact on the quality and taste of foods produced. Adjustment of fertilisation to a plant species, and sometimes even to a cultivar, has an enormous effect on plant health and growth as well as on yield quality and quantity. Nitrogen fertilisation requires high precision because nitrogen not bound by plants is usually lost due to its being leached to groundwater or volatilised into the air. Nitrogen application rate and timing require great caution. Nitrogen rates should be adapted to the requirements of a cultivar grown, weather conditions and expected yield. An excess of nitrogen is the cause of accumulation of nitrates in some vegetable species such as, e.g., zucchini, carrot, lettuce, spinach and beetroot, whereas an excess of nitrates ingested with plants is harmful to people [Abdussalam-Mohammeda 2020; Anastas & Eghbali 2010].

Sustainable farming is society's main need that can be satisfied thanks to different production models, including organic agriculture and integrated agriculture termed alternative agriculture. An alternative to industrial farming is organic and sustainable farming. Organic methods are considered to be environmentally friendly because such methods do not use chemisation and they cooperate with nature. An agricultural production method and environmental conditions may affect the quality of food raw materials. Organically produced food has many beneficial nutritional, health and sensory characteristics compared to food from conventional farms. It contains less pesticide residues as well as nitrates and nitrites, but more vitamins, valuable protein and minerals. What is more, it has better taste and smell qualities, which is important to consumers. Compared to conventional farming, alternative agriculture does not seek to obtain high yield at the cost of high expenditures on agricultural chemisation. The aim of alternative agriculture is to produce food in a way that does not degrade the natural environment [Fenibo 2021].

3.5 OTHER CHEMICAL HAZARDS IN AGRICULTURAL PRODUCTION AND FOOD PRODUCTION

Chemisation of the natural environment and the related common occurrence of hazardous and harmful chemical substances of different types are indispensably associated with the occurrence of these substances in food. Potential chemical contaminants in food can be divided into a number of groups:

1. Natural toxins: mycotoxins, sea fauna biotoxins;
2. Environmental contaminants such as, e.g., heavy metals, dioxins, furans, and radionuclides;
3. Chemicals used in food processing that may get into food as a result of the leakage of lubricants, machinery damage, and residues of cleaning agents and disinfectants;
4. Compounds and polymers migrating from materials in contact with food;
5. Residues of agri-chemical procedures, veterinary drugs;
6. Toxicants produced during processing, in particular in the course of thermal processes: acrylamide, furan, PAHs (polycyclic aromatic hydrocarbons), HAAs (heterocyclic aromatic amines) and others [Nicopolou-Stamati *et al.*, 2016].

Issues related to contamination of the natural environment and agri-food products arouse an ever greater interest in society. As a result of education conducted in mass media, in particular with regard to health hazards caused by consumption of microbiologically or chemically contaminated food, not only functional, sensory, hygienic and aesthetic properties are now taken into consideration in consumers' assessment of agri-food products, but also threats from foreign substances, xenobiotics, in a product that may present a risk to health safety and well-being [Rani *et al.*, 2021].

Food safety is an integral part of the safety of the entire agri-food economy. The lack of food safety has a negative impact on the condition of a national economy and also households by closing export markets or reducing the availability of food in the domestic market. The consumer has the right to safe food that will not cause a threat to his health and life, that is, his welfare. Reduced consumption of mass produced foods in favour of regional, traditional and organic products can be observed among consumers, especially those more affluent ones [Witczak & Abdel-Gawad 2014].

The EU's common agricultural policy, whose aim is not only to provide a sufficient amount of food, but also high quality food produced in a sustainable way in keeping with the requirements regarding protection of the environment, water resources, animal health and welfare, plant health and public health, responds to consumers' expectations. In the report "*Comparison of composition (nutrients and other substances) of organically and conventionally produced foodstuffs*", a review of 162 scientific publications was conducted, the research methods applied and the quality of the studies performed were verified, and based on this review the following conclusions regarding food quality were presented:

1. As far as conventional and organic food of plant origin is concerned, no differences were found as regards the content of vitamin C, calcium, phosphorus, potassium, copper, iron, nitrates, magnesium, ash, specific proteins, sodium, non-digestible carbohydrates, beta-carotene and sulphur as well as titratable acidity.
2. Significant differences were found in the content of minerals in plant derived foods (a higher nitrogen content in conventionally produced foodstuffs, whereas a higher magnesium and zinc content – in organic products), phytocompounds (a higher content of phenolics and flavonoids in organic foods) and sugars (a higher content in organic foods).
3. As regards organically and conventionally produced foods of animal origin, no differences were found in the content of nutrients and other substances such as saturated fatty acids, cis-monounsaturated fatty acids, omega-6 and omega-3 polyunsaturated fatty acids, nitrogen and ash [Dangour *et al.*, 2009].

Over the period 2002–2011; the State of Baden-Wurttemberg conducted monitoring of the quality and safety of organic foods under a federal-wide food control programme. Based on the results of this research regarding chemical contamination, the findings were as follows:

1. Organic fruits and vegetables generally contained lower pesticide concentrations compared to conventionally produced ones. At the beginning of the monitoring period, antibiotic residues were detected in about 25% of honey samples – now they are detected rarely.
2. Products subjected to heat processes were interesting. In the case of organic coffee that was subjected to the roasting process, a significant concentration of furan was found, whereas in the case of organic crackers and potato chips the acrylamide content was determined to be above the signal value, i.e. 1000 µg/kg.
3. In food of animal origin, eggs from free-range hens and beef meat, the contents of dioxins and dioxin-like compounds were much higher than for organic products (high above the maximum residue limit, MRL). This was due to the fact that hens were kept outside (free-range).
4. The most frequently investigated contaminants in organic food are mycotoxins, metabolites of microscopic fungi colloquially termed mould, which exhibit strongly toxic effects towards human organism and animals.

Contamination of cereal grain with mycotoxins is an important problem of modern farming in many countries. Their harmful effects are manifested at low concentrations – at a level of about 1 milligram per kilogram, i.e. the one millionth part of the grain weight or even less. Their ingestion with animal feed causes mycotoxicoses – diseases characterised by a specific effect in livestock, such as liver damage and tumours in the case of aflatoxins, kidney damage in the case of ochratoxins A, fertility disorders in pigs due to zearalenone as well as loss of appetite and vomiting in pigs after ingestion of deoxynivalenol (vomitoxin). At the same time, all mycotoxins exhibit non-specific effects manifested in reduced utilisation of feed and overall health deterioration in animals. Fusarium mycotoxins are the greatest problem. Their concentrations in conventional and organic foods are at a similar level with a declining trend for processed products [Kwasek & Obiedzińska 2014].

A significant environmental polluter is animal husbandry, which is responsible for 18% of global greenhouse gas (GHG) emissions from human activity, measured in CO_2 equivalent. It is more than the percentage of transport, responsible for 14% of global GHG emissions. Most of these 18% are emissions of nitrogen suboxide and methane originating from animal manure as well as emissions of methane from animal digestive processes and of nitrogen suboxide from mineral fertilisers used in animal feed crops. The animal husbandry sector's percentage share in anthropogenically-induced global primary GHG emissions is as follows: 37% of total methane emissions, 65% of nitrogen suboxide emissions, and 9% of carbon dioxide emissions. 64% of emissions of ammonia, which contributes to air, soil and water pollution, acid rain formation, and ozone layer damage, also originate from livestock production [Li *et al.*, 2021; Shi *et al.*, 2022; Yang *et al.*, 2019].

Chapter 4

Food production and processing

4.1 INTRODUCTION

Due to progress in agriculture that has been made over the last decades, food production has outpaced the increase in population. The common use of artificial fertilisers, plant protection products, antibiotics and other drugs in animal husbandry significantly increases agricultural production efficiency, but they also have an impact on the quality of products obtained, which are a raw material base in the food industry. The method of processing agricultural commodities to produce foods and additives used for this purpose, which allow the chemical composition and physical properties to be modified, improve taste or smell qualities, enable longer storage, etc., are important factors that determine food quality.

Consumers' awareness of food related risks is continually growing and this contributes to a gradual change in people's eating habits. They more and more often pay attention to the safety of food, its origin, or the type of additives used.

This chapter discusses problems associated with (organic and non-organic) food production and processing, with special reference to raw materials of plant origin. It presents the outstanding role of plant products in the current approach to the food pyramid. An analysis is also made of food parameters that affect food quality assessment. The relationship between food components and the nutritional value of a product is highlighted. The chapter finishes with issues regarding biological, chemical and physical risks during food production and storage.

4.2 FOOD PRODUCTION

In recent years, the increase in food production has outpaced the increase in population. This has occurred primarily due to progress in agriculture as well as the use of various crop cultivars with greater productivity (short-stem wheat and rice), mineral fertilisers, and plant protection products (more effective pest control). Owing to a balanced feeding system and the introduction of new animal breeds, the productivity of animal husbandry has increased. The greatest achievements in agricultural production occurred after World War II.

Production of new raw materials with a modified chemical composition, thanks to application of biotechnology and genetic engineering, has led to changes in the genetic code of plants and animals. In the long term, we cannot foresee consequences of these changes, both beneficial and adverse ones. The transfer of genes between plant species may lead to

DOI: 10.1201/9781003380771-4

production of products with altered sensory characteristics (taste, smell) and also containing allergens. For example, the replacement of old apple varieties with other ones resistant to diseases and pests has led to production of fruits with worse taste qualities.

Rapeseed oil belongs to plant raw materials with a modified composition. The old high erucic acid and high glucosinolate rapeseed varieties were characterised by a high erucic acid content (about 50%) and low sensory qualities. The modern improved low erucic acid or erucic acid free rapeseed varieties have a favourable fatty acid composition, desirable from the point of view of health, the presence of active ingredients, tocopherols and phytosterols, as well as a neutral taste and smell [Buiatti *et al.*, 2013; Kramkowska *et al.*, 2013; Kubisz *et al.*, 2021; Raman 2017].

Sustainable and organic farming is an alternative to industrial agriculture. The research results reveal that environmental conditions and agricultural production methods may affect the quality of food raw materials, at the same time ensuring biodiversity of the environment. It has been shown that food from organic farms has many beneficial health, nutritional and sensory characteristics compared to conventionally produced food. It contains less nitrates and nitrites as well as less pesticide residues, but more vitamin C and other vitamins, total sugars, valuable protein, and minerals. Apart from that, it has better flavour and aroma qualities, which is of significance for consumers [Hallmann & Remproteinwska 2007; Golijan & Sečanski 2021; Mukherjee *et al.*, 2020]. Drying, particularly freezing that inhibits chemical and biological processes due to reduced temperature of a raw material, belongs to the oldest and most effective food preservation methods [Haiying *et al.*, 2007].

One of the first forms of "convenience food" was cereal coffee. In Poland the first factory that manufactured cereal coffee was set up in 1816. The greatest development of the food industry has taken place since the 19th century. Industrial-scale food preservation, in order to meet needs of large social groups, developed at the turn of the 18th and 19th centuries thanks to application of sterilisation technology invented by Appert. An important stage in the development of this type of foods was the launch of industrial-scale production of bullion products from meat extracts by Gilbert in Uruguay in 1864. In 1880 in Switzerland, Maggi founded a factory producing seasonings, the so-called bouillon cubes produced using the method of protein acid hydrolysis. This laid foundations for the development of the future food concentrate industry [Janicki 1993]. The need to feed soldiers under battlefield conditions in the 19th and 20th centuries was an impulse for the development of the food concentrate industry. Production of highly concentrated food with an appropriate nutritional value and sensory quality required the development of science and technology. During World War II, there was an improvement in the production of food concentrates and tinned preserved food. Thanks to enhanced technology, the quality of products improved. At the end of the 20th century, there was a further development of "convenience food" [Świderski 2003].

The processing industry uses various functional additives: preservatives, stabilisers, colourings, flavourings, etc. Raw materials are "enhanced" by removing certain substances and adding other ones (e.g. yogurt enhanced with live bacterial cultures, coffee with magnesium, juice enhanced with vitamins, calcium and dietary fibre). Vitamins and minerals are added to functional food to prevent or correct nutrient deficiency in whole populations or specific population groups, but also to replenish losses that occurred during food processing and storage [Ratkovska *et al.*, 2007]. Due to the continuous availability of food, an increasingly larger group of consumers expect food with attractive sensory properties and valuable in health terms.

In the last decade of the 20th century, health-promoting foods appeared whose composition is designed to support the organism in maintaining physical and mental well-being and to prevent or treat diseases. In its composition, functional food has groups of compounds that affect its biological activity: dietary fibre, oligosaccharides, polyunsaturated fatty acids, amino acids, peptides, proteins, glycosides, vitamins, minerals, alcohols and phenols, lactic acid bacteria, choline and lecithin, phytochemical substances, etc. Among this food, the following products are distinguished: fortified, low energy, high fibre food for people with specific physiological conditions (allergy sufferers, diabetics, infants, and pregnant women), food reducing the risk of non-communicable diseases (e.g. low cholesterol food, food with a reduced sodium content, probiotic food) and food for athletes [Butinariu & Sarac 2019; Drobnica *et al.*, 2007; Henry 2010].

Efforts are made to encourage consumers to eat low processed food by using attractive packaging and terms such as: "novel food", "identical with natural", "according to grand-mother's recipe" or "according to regional recipe", "exclusively natural ingredients", "without preservatives", "tradition of taste" or "non-GMO". Traditional processing methods are still used: pasteurisation, homogenisation, concentration, stabilisation, drying, thickening, lique-faction and thinning. However, some of them has lost importance, for instance salting. Not all changes in industrial food production should be perceived as negative; for example the microbiological quality of milk in the 1960's and 70's was at a distinctly lower level than nowadays. Industry's interest in innovations began in the 1940–50's. Among innovative products, we can distinguish the following food products: with flavouring additives, with new flavours, "healthier" products containing new ingredients, including bioactive ones, specia-lised products (e.g. for allergy sufferers, gluten free), for specific groups of consumers (e.g. intended for males), convenience food, products produced according to an old recipe or products in new packaging [Kowalczuk 2011; Temple 2022].

Cereal foods, including snacks, are often carriers of bioactive substances in food production because foodstuffs that are regularly purchased and consumed by consumers are chosen. Such food products may include corn puffs with the addition of natural organic ingredients enriching their chemical composition, such as amaranth, Jerusalem artichoke or pumpkin. These ingredients are interesting both to nutrition technologists and specialists due to their rich nutrient content. At the same time, this highly processed (extrusion) product, i.e. corn puffs, is a frequently consumed snack that predominantly meets hedonistic needs, not health ones. Enriching this product with nutrients can be an impulse to purchase it for consumers whose healthy eating awareness is increasing [Kremen & Miles 2012; Kwiatkowski & Harasim 2019].

4.3 THE ROLE OF PLANT PRODUCTS IN THE PRESENT APPROACH TO THE FOOD PYRAMID

The current trends in consumption of particular foods aim at popularising plant diets, which include vegetarianism and veganism. Plant diet is a nutrition trend where foods of animal origin, such as eggs, meat, meat products, milk and dairy products, are sig-nificantly reduced or completely eliminated. It is assumed that diet performs an important role not only in the aspect of physical health, but also in the proper func-tioning of the nervous system – contributing to the development of cognitive and mental health. Diets based on consumption of plant foods are considered to be consistent with the latest guidelines of the World Health Organisation (WHO) regarding public health preservation. As shown by one of the world' most detailed research studies in this respect, *The European Prospective Investigation into Cancer and Nutrition*, plant-based diets may contribute to decreasing the risk of cardiovascular diseases due to a significant reduction in consumption of saturated fats. Moreover, they can minimise the danger of overweight, obesity, digestive system conditions, and even tumours – especially stomach cancer and lower gastric cancer [Fehér *et al.*, 2020; Iguacel *et al.*, 2020].

One of the reasons for such an abrupt interest in plant-based diets in recent years is undoubtedly the perception of the scale of the problem resulting from climate change. They have become the subject of deliberations of many consumers who want to minimise the existing environmental problem. It is worth indicating that total food production – including processing, distribution and consumption – contributes to the deepening of many climate transformations, for example reduction in freshwater resources, water, soil and air pollution, and in consequence also food pollution and decrease in biological diversity. This has led to a situation when the particular food production systems have become the essence of climate change mitigation. Taking into account greenhouse gas emissions, being a real problem, the difference between production of animal origin food and plant origin food is considered to be significant. Specialists in this subject raise the issue of plant diet as a tool designed to reduce individuals' impact on the planet [Kortetmäki & Oksanen 2020]. Given the above, appropriate recommendations containing nutrition related guidelines for sustainable consumption should be developed. In many countries, public institutions address the topic of sustainable diets, among others in Sweden, the United Kingdom and the Netherlands. These countries conduct research regarding sustainable consumption and sustainable diet. The impact of one's diet on greenhouse gas emissions was the subject of a study performed by F. Vieux and other researchers on the example of a population of French adults. This study reveals that high-nutritional-quality diets have a greater effect on the environment than low-nutritional-quality ones (the GHG values were higher by +9% and +22% for men and women, respectively). Furthermore, this research confirms that per 100 g of product, food of animal origin (ruminant meat, pork, poultry and eggs) shows higher GHG emission values than plant-based foods [Vieux *et al.*, 2013].

Sweden is the first country in Europe that has proposed a change in the modern food consumption pattern in order to make it safe both for humans (as regards rational diet) and for the environment (in terms of the level of pollution and greenhouse gas emissions in food production and marketing as well as in terms of the use of chemicals in agricultural production). The Swedish National Food Administration (NFA) developed recommendations relating to six groups of food products: meat and processed meat, fish and shellfish, fruits and vegetables, potatoes, cereals (including rice) as well as fat and water. A study was also carried out in Sweden which compared balanced meals (in nutritional terms) consisting of local and imported foods. This study shows that a vegetarian meal consisting of local foods has a nine times lower Global Warming Potential (GWP) than a meal containing pork and imported products [BCFN 2012a]

The Barilla Center for Food & Nutrition presented a double pyramid: the food pyramid and the environmental pyramid. The food pyramid was built based on the nutritional properties of food products. In the environmental pyramid, foods were classified based on their impact on the natural environment. As a result, an inverted pyramid was obtained relative to the food pyramid: at the top of the pyramid there are foods that have the greatest impact on the environment, whereas at the bottom of the pyramid – those of lesser importance. The placement of these two pyramids alongside illustrates that food with higher recommended consumption levels (e.g. vegetables and fruits recommended to be eaten five times a day) have the smallest impact on the environment, whereas foods whose consumption should be reduced (e.g. red meat) have the greatest effect on the environment. The double pyramid has two important goals: to preserve human health and to protect the natural environment. In other words, food that beneficially affects human health has at the same time a positive effect on the environment. Keeping a balanced diet generates lower environmental costs (Carbon Footprint, Water Footprint), but it also contributes to reduction in health costs related to diet-dependent diseases. Sustainability requires a continuous balance over time at several levels: environmental, social and economic. Different eating choices of consumers influence their food expenses. The price of a food product is affected by its origin (e.g. products of animal or plant origin), quality, the place of sale (hypermarket, supermarket, retail shop), the geographic region and the seasonality [BCFN 2012b].

Another reason that determines complete or partial abandonment of meat and/or animal-based foods is an attempt to prevent avoidable suffering of animals from which such products are obtained. Therefore, the new nutrition models developed over the several last years are largely driven by ethical considerations [Kortetmäki & Oksanen 2020].

The new interpretation of the food pyramid, or consumers' diverse approach to it, arises from different lifestyles of particular societies, which are influenced, among others, by cultural, professional and financial considerations as well as by convictions and values held. Generally speaking, a consumer's choice of diet is affected by the widely understood habitus of an individual. The society, going a step further towards an ethical lifestyle, attempts to undertake environmental protection activities with deliberation by, *inter alia*, choosing organic food or preferring meat-free diets. It should also be noted that globalisation processes are a tool used to popularise certain behaviour patterns, including a change in consumer's eating preferences and following these changes by food providers [Mamzer 2020].

The consequences of changes in eating habits in particular societies have been observed over the last several decades. It was during this time that there was a breakthrough in the food industry that faced new challenges – attention was focused, among others, on the production of new types of food that could reduce to some degree the prevalence of some chronic diseases. Diets that increasingly attracted interest, and therefore also diets based mainly on plant products, began to be analysed. Numerous questions concerning not only environmental considerations, but also health issues appeared at that time. Broad research was conducted, during which objective benefits and benefits perceived by consumers arising from the choice of a plant-based diet were described. Over the years, it can be noticed that society, by giving up consumption of meat and/or animal-based foods, redefines the meaning of convenience and taste factors, which were formerly associated to a greater degree with traditional diets [Fehér *et al.*, 2020].

Agriculture is the starting point for understanding a sustainable diet where already the choice of the system of production of agricultural commodities, which are plant raw materials for further processing, affects diet components as well as the economic, environmental, health and cultural factors. A sustainable diet must therefore result from the activities of a sustainable food chain based on sustainable agricultural activity [Allen 2013]. Organic production has sense only when the entire agri-food chain implementing practical organic production principles is preserved, whereas the guidelines contained in the legislation are reflected in good agricultural production and food manufacturing practices – which may form the basis for considering such practices to be "green technologies". In other words, production, storage, transport and processing methods must be consistent with the principle of oversight and compliance "from field to table", as specified in the food law requirements [Kahl *et al.*, 2010]. The organic production method performs a double social function: on the one hand, it provides commodities to a specific market driven by the demand for organic products, while on the other hand – it is a public interest activity because it contributes to environmental protection, animal welfare and rural development. Currently, many countries lack a specialised system for monitoring chemical contamination and residues (chemical hazards) as well as critical quality indicators and traceability, which would cover the entire organic food chain in temporal and spatial terms in order to support the reliability of health and safety statements related to organic food and its origin. To ensure consumers' health and economic safety, solutions should be developed for systemic food defence, i.e. protection of foods against intended contamination with biological, chemical, physical or radio-nuclear agents in the entire agri-food chain – "from field to table". A necessary element of such a solution is to establish a specialised scientific and research centre that would support the provision of quality and safety in organic farming and organic food. Each country's food policy should seek to harmonise different activities (policies) with society's food and health needs. It must shape consumers' attitudes in such a way that they choose appropriate food consumption patterns (proper nutrition, rationalisation of nutrition). An important task of food policy is also to conduct elementary food and health education, both among adults and children [Bach-Faig *et al.*, 2011; Kwiatkowski & Harasim 2019].

4.4 PROCESSING OF ORGANIC AND NON-ORGANIC FOOD OF PLANT ORIGIN

Processing of plant products, which have an ever greater importance in diet, is continuously tailored to consumers' needs and expectations with regard to the final product. An increasingly larger proportion of slightly processed products, characterised by certain properties desirable for health, or products manufactured based on strict quality guidelines, such as organic food or other foods with specific designations, can be noticed in recent years. Despite that consumers' knowledge concerning food related risks is gradually growing and a change in eating habits to healthier ones is becoming an important issue, society's current eating habits are still largely based on industrially processed products containing various additives in its composition. Their use means an interference with a product's composition that will enable a technological process to proceed and/or to be significantly accelerated, which is the main benefit for its producer. Food additives are also substances that allow a wider range of products to be created by modifying their composition and giving them new characteristics. Moreover, these additives perform a vital role in developing repeatable quality of food – characterised by a constant colour or flavour [Kotynia & Szewczyk 2019; Mie *et al.*, 2017].

Additives sometimes do not have at all or have an only insignificant nutritional value, but apart from increasing the attractiveness of a product, their use in the food industry is designed to extend its shelf life. Many food preservation methods slowing down adverse processes that occur in the finished product, which include preservation in vinegar and salting, were already known in the past centuries. However, as a result of consumers' pressure and the fast growing food industry, producers started to add newer and newer additives that allowed them to meet buyers' expectations, thus creating competitive goods [Pandey & Upadhyay 2012].

The difficulties with incorporating foods with a shorter, simpler composition into people's diet are deepening with processors' growing orientation towards financial benefits arising from their efforts to achieve processability of a raw material and undisturbed processing, which is possible due to using additives [Gruchelski & Niemczyk 2019]. These substances do not occur as stand-alone food products but only as auxiliary agents for which the labelling system is uniform across the entire European Union. Designations consisting of the "E" symbol and a relevant number, which indicates their assignment to a specific group of additives performing the same technological function, are assigned to them. The most important aspect in processing both organic and conventional food is the widely understood food safety. Certification bodies perform an extremely significant function because thanks to them it is possible to completely eliminate or minimise hazards that may appear at each stage of organic food production, processing and distribution, and this function arises from the need to build trust between producers and consumers for whom quality as well as high taste and nutritional values are of priority importance. Control of organic food is also designed to exclude marketing of fake food products with misleading declarations on their labels [Golijan & Sečanski 2021].

In processing of conventional food, 330 additives with different applications are permitted. Depending on the assumed specificity of the finished product – they can perform as many as 27 different functions in relation to processes conducted. The general division of food additives is as follows:

a) Food colours (E100 – E199);
b) Preservatives (E200 – E299);
c) Antioxidants and acidity regulators (E300 – E399);
d) Thickening agents and emulsifiers (E400 – E499);
e) Processing aids (E500 – E599);
f) Flavour enhancers (E600 – E699);
g) Glazing agents, sweeteners, etc. (E900 – E999);
h) Thickeners, stabilisers, etc. (E1000 – E1999).

Before being permitted for use, each additive was tested for food safety. The maximum limits that do not cause adverse effects on health were also determined or the "quantum satis" principle is applied, which recommends the use of an amount of a specific additive which is necessary to carry out a given technological process. The European Food Safety Agency (EFSA) is an agency that deals with analysis of potential harmfulness of these substances and it regularly prepares and updates reports on additives in processed products. Despite that all currently used compounds have been approved for consumption at specific safe levels, there is no unambiguous information whether simultaneous consumption of many food additives, in particular from different groups, will result in a threat to consumers' health, or not. This primarily relates to the possibility of inducing synergistic effects of two or more compounds and the impact of synthetic additives on natural ingredients occurring in food [Chazelas et al., 2020].

Organic products, in the case of which only selected additives are permitted for use in processing processes, are an alternative to conventional foods. An organic food is a product where organic ingredients were used in at least 95% to produce it. The other ingredients, also including food additives, may originate from conventional agricultural production in some cases. Nevertheless, a condition for its processor is to demonstrate that their use is necessary to carry out the technological process or there are no equivalent organic raw materials. Furthermore, one should ensure process cleanliness, thus separation in time, and – if possible – the use of separate production lines if a given manufacturing facility is involved in both conventional and organic food processing [Sharma 2017].

The use of additives in processing organic plant products (Table 4.1) and the permissible processing methods are governed in EU countries by the following regulations: Regulation (EU) 2018/848 of the European Parliament and of the Council of 30 May 2018 on organic production and labelling of organic products and repealing Council Regulation (EC) No 834/2007, 2018, and Commission Implementing Regulation (EU) 2021/1165 of 15 July 2021 authorising certain products and substances for use in organic production and establishing their lists. The most important issues raised in them relate to the principles that allow the natural characteristics of a raw material to be preserved and therefore they apply to processing that does not significantly affect the original physico-chemical properties of foods. It is possible thanks to the use of physical methods or fermentation processes that ensure obtaining an only slightly changed product in comparison with some processed conventional products.

Table 4.1 Selected additives and processing aids authorised for use in the production of processed organic food [on the base of Commission Implementing Regulation (EU) 2021/1165].

Name and cod	Foodstuff to which it can be added / Specific conditions and limits	Name and cod	Foodstuff to which it can be added / Specific conditions and limits
Calcium carbonate (E 170)	Plant and animal origin products / Not be used for colouring and enrichment the products with calcium	Guar gum (E 412)	Plant and animal origin products / only from organic production
Sulphur dioxide (E 220)	Wines made from fruits other than grapes and mead / Max. 100 mg SO_2 /dm^3	Arabic gum (E 414)	Plant and animal origin products / only from organic production
Potassium metabisulphite (E 224)	Wines made from fruits other than grapes and mead / Max. 100 mg SO_2 /dm^3	Xanthan gum (E 415)	Plant and animal origin products / No limits
Lactic acid (E 270)	Plant and animal origin products / No limits	Gellan gum (E 418)	Plant and animal origin products / Only high-acyl form from organic production
Carbon dioxide (E 290)	Plant and animal origin products / No limits	Glycerol (E 422)	Plant extracts flavouring / Only plant origin and from organic production
Malic acid (E 296)	Plant origin products/No limits	Pectin (E 440i)	Plant origin and milk-based product/ No limits
Ascorbic acid (E 300)	Plant origin and meat products / No limits	Hydroxypropyl methyl cellulose (E 464)	Plant and animal origin products / Encapsulation material for capsules
Tocopherol-rich extract (E 306)	Plant and animal origin products / Antioxidant	Sodium carbonate (E 500)	Plant and animal origin products / No limits
Lecithins (E 322)	Plant origin and milk products / only from organic production	Potassium Carbonates (E 501)	Plant and animal origin products / No limits

(continued)

Table 4.1   Continued

Name and cod	Foodstuff to which it can be added / Specific conditions and limits	Name and cod	Foodstuff to which it can be added / Specific conditions and limits
Citric acid (E 330)	Plant and animal origin products / No limits	Ammonium carbonates (E 503)	Plant and animal origin products / No limits
Sodium citrates (E 331)	Plant and animal origin products / No limits	Magnesium carbonates (E 504)	Plant and animal origin products / No limits
Calcium citrates (E 333)	Plant origin products / No limits	Calcium sulphate (E 516)	Plant origin products / carrier
Tartaric acid (E 334)	Plant origin products and mead / No limits	Sodium hydroxide (E 524)	Lye rolls (Laugengebäck") flavourings / Acidity regulator and surface treatment agent
Sodium tartrates (E 335)	Plant origin products / No limits	Silicon dioxide (E 551)	Herbs and spices in dried powdered form, flavourings and propolis / No limits
Potassium tartrates (E 335)		talc (E 553b)	Sausages based on meat / Surface treatment
Monocalcium phosphate (E 341i)	Self-raising flour / Raising agent	Beeswax (E 901)	confectionery / only from organic production used as glazing agent
Extracts of Rosemary (E 392)	Plant and animal origin products / Only from organic production	Carnauba wax (E 903)	Confectionery citrus fruits / only from organic production, used as glazing agent, methods for mitigating extreme cold, as a quarantine measure against harmful organisms
Alginic acid (E 400)	Plant origin and milk products / No limits	Argon (E 938)	Products of plant and animal origin / No limits
Sodium alginate (E 401)	Plant origin, milk products and meat-based sausages / No limits	Hel (E 939)	Products of plant and animal origin / No limits
Potassium alginate (E 402)	Plant origin and milk products / No limits	Azot (E 941)	Products of plant and animal origin / No limits
Agar (E 406)	Plant origin, milk-based and meat product / No limits	Tlen (E 948)	Products of plant and animal origin / No limits
Karagen (E 407)	Plant origin and milk-based products / No limits	Erythritol (E 961)	Plant and animal origin products / only from organic production without using ion exchange technology
Locust bean gum (E 410)	Plant and animal origin products / Only from organic production		

The organic food processing method should only be limited to processes necessary to obtain the desired product, thereby not losing its valuable original features. It is prohibited to use genetically modified organisms, isolated ingredients, and radiation methods of food preservation – ionising irradiation that affects a number of biological and chemical transformations in the final product. It is also prohibited to use treatments designed to restore the characteristics of organic food lost during the technological process [Średnicka et al., 2009].

4.5   ANALYSIS OF FOOD PARAMETERS AS FOOD QUALITY ASSESSMENT

Testing of food technology and safety conducted by representatives of the food industry often requires an accurate analysis of the composition and properties of foods. Any conditions set by consumers force certain actions and also pose a challenge to scientists in the context of monitoring and ensuring the quality and safety of products. With the increasing availability of high capacity methodology used to characterise and analyse food, more and more data on foods are gathered which can be used to certify their quality. Product quality management in the food industry has an ever greater importance and hence analytical methods must be applied in the entire food supply chain in order to achieve the required quality of the final product. Selection of individual methods is usually determined by the nature of a sample tested and the specific reason for performing an analysis. Individual tests are unique for each ingredient or properties that are to be checked [Biancolillo 2019; Nielsen 2009].

Food analysis is a process employed for both fresh foods and products processed according to standard procedures, and such analysis is used to provide information about many different characteristics of food, including its composition, structure, physico-chemical properties and microbiological characteristics. Quality control is the maintenance of quality at a constant level and within tolerance limits acceptable to the buyer, while minimising costs for the seller. The quality and safety of foods depend on the percentage content of carbohydrates, protein, fat, dietary fibre, vitamins or minerals as well as on hidden attributes such as peroxides and free fatty acids. This information is of key significance to rationally understand factors that determine food properties and also to be able to economically produce food that is consistently safe, nutritional and desired as well as for consumers making conscious choices regarding their diet. As a matter of fact, the most important element and ultimate goal in food quality analysis is to protect the consumer and to provide to him reliable information on the product purchased [Rajput & Rehal 2019].

4.6 BASIC FOOD COMPONENTS AFFECTING THE NUTRITIONAL VALUE OF A PRODUCT

At the stage of production, processing and distribution, food is exposed not only to pollution, but also to the loss of valuable nutrients and therefore it is important to cyclically monitor its composition. The starting point for determining the basic nutritional value of a raw material or the finished product is to determine its water content.

The water content in plant products depends on many factors and is different for individual types of food. Its percentage proportion in fruits and vegetables not subjected to technological treatment may range from 75 to even 95%, while in the case of legumes these values reach about 10%. The content of this component in food is determined by the storage method and duration and hence it is not a constant parameter. Water in food performs an essential function of regulating physico-chemical transformations and supporting enzymatic and microbiological processes. It also contributes to the development of sensory characteristics of a product – among others its texture and rheological properties. Water is the most abundant component of food, whereas from the point of view of food safety, it is the most significant component. Its presence, amount and character determine many chemical and biochemical processes important to control product safety and quality. The more water in a product, the smaller the percentage of dry matter is; this is adverse from the point of view of food storage because food with a high water content is stored much worse and moreover it is associated with a lower percentage of desirable food components. From the analytical point of view, the term "dry matter" means the remaining components in a product or raw material after water has been completely removed from it [Rahman and Labuza 2020; Zhang *et al.*, 2015].

Other important food components include protein, which is a source of energy, alongside carbohydrates and lipids. Furthermore, proteins perform a number of other functions in the organism, such as increasing enzymatic activity and transporting nutrients and other biochemical compounds through cellular membranes. To maintain these functions, it is necessary to provide to the organism good quality proteins through diet. Due to the fact that protein is an important component of human diet, it is essential to know its content in food and therefore to have reliable analytical methods. Analysis of food protein is not always a simple procedure. This is partly attributable to the fact that food is a heterogeneous material consisting of a range of different nutrients such as lipids, carbohydrates and various micronutrients. The composition, structure or matrix of food as well as interactions between different nutrients may reduce the availability of proteins, which leads to underestimation of protein content. Furthermore, different methods are based on diverse analytical principles, determining protein content in a direct or indirect way. Direct determination of protein content is when the protein content is calculated based on the analysis of amino acid residues [Mæhre et al., 2017].

Carbohydrates – macronutrients that predominantly occur in plant products and are found in fruits, grains and vegetables – should also be mentioned among basic food components. Most saccharides of plant origin are found in cereal products – it is accepted that their percentage is from 50 to 80%. These organic compounds are present in food in the form of sugars, starch and fibres, and they are composed of carbon, hydrogen and oxygen. They are widespread and according to their chemical structure they can be classified into three major groups:

1) low molecular weight mono- and disaccharides;
2) intermediate molecular weight oligosaccharides;
3) high molecular weight polysaccharides.

It is worth noting that the digestibility of carbohydrates varies and non-digestible carbohydrates may reach the large intestine intact, where they are a source of food for beneficial bacteria [Kiely & Hickey 2022].

The advantages of plant products include the occurrence in them of large amounts of dietary fibre (roughage) resistant to the action of digestive enzymes – due to resistant starch. Dietary fibres are the major component of low energy products that have gained an increasing importance in recent years. They also have specific technological and functional properties that can be used in the formulation of foods as well as numerous beneficial effects on human health. The current state of knowledge shows that dietary fibre may contribute to a reduced risk of many diseases, including duodenal ulcer, obesity, diabetes, cardiovascular diseases as well as stomach and colon cancer [Ötles & Ozgoz 2014].

Lipids, whose mixtures make up compounds called fats, are a large group among food components. Lipids should provide about 50% of daily intake of energy in human diet. In terms of their chemical structure, the following lipids can be distinguished:

1) simple lipids (proper lipids and waxes);
2) complex lipids (phospholipids and glycolipids);
3) secondary lipids (fatty acids, lipid alcohols and hydrocarbons).

Apart from their energy function, fats also perform the role of carriers of substances important for the proper functioning of the organism, such as vitamins A, D, E and K, essential unsaturated fatty acids or steroids. In addition to their effects on human organism, they are a factor that determines some technological processes, which include controlled heat transfer. They also determine rheological properties of products and can be used as emulsifiers, for instance lecithin. Raw materials that are abundant in valuable fatty acids include nuts as well as seeds or fruits of oilseed plants such as rapeseed, sunflower, flax and sea buckthorn. Cold pressed oils, which can be obtained through mechanical processes without using high temperature in their production, are particularly desirable. Oil obtained in this way is characterised by a high content of mono-unsaturated and polyunsaturated fatty acids that have an impact on the proper functioning of the circulatory system and metabolic transformations in the organism [Obiedzińska & Waszkiewicz-Robak 2012].

Plant raw materials are a source of antioxidants, colouring substances, vitamins and minerals. Phytochemical compounds with health-enhancing properties are used, among others, in the production of nutraceuticals designed to maintain certain physiological processes, thereby preventing some diseases. They perform regulating functions and are only used in specific cases, among which prevention and treatment of diseases resulting from deficiencies caused by improper nutrition should be mentioned. Vitamins and minerals taken with plant food that are desirable in diet include carotenoids, vitamin C, selenium, magnesium, potassium, zinc, iron, tocopherols, polyphenols and many others [Chmielewska *et al.*, 2018].

4.7 BIOLOGICAL, CHEMICAL AND PHYSICAL HAZARDS DURING FOOD PRODUCTION AND STORAGE

The development of the economy has a significant impact on increasing the number of hazards in the aspect of food safety. Production of safe foods requires accurate control, which is enabled by food quality and safety management systems. These systems involve control of food products at each stage of their production, that is, from the producer to the consumer [Hussain *et al.*, 2017; Sitarz & Janczar-Smuga 2012].

Among biological food hazards, microorganisms are an essential threat. Food poisoning and food-borne infections, caused mainly by the following bacteria: *Salmonella, Escherichia coli, Campylobacter*, and *Listeria monocytogenes*, have been the most frequent problem for many years. The most common viruses causing digestive system conditions are rotaviruses and the hepatitis A virus. Fungi of the genera Penicilium, Aspergillus, and Fusarium are also a source of infections. Internal parasites, such as *Entamoeba histolytica, Giardia lamblia, Toxoplasma gondii*, cestodes, nematodes and trematodes make up another group. From the sanitary point of view, these hazards also include rodents and insects, which are a danger in themselves and also a habitat for microbes [Schirone *et al.*, 2017; Visciano *et al.*, 2016].

Chemical hazards include all substances that got into food at any stage of the food chain. These contaminants seriously affect food quality and safety. They result in multitudes of poisoning cases, diseases and adverse changes in human organism that may manifest themselves in the next generation, e.g. in the form of cancer, congenital defects, fertility disorders or immune and nervous system damage [Hussain *et al.*, 2017; Sitarz & Janczar-Smuga 2012]. The most common potential chemical hazards in food are the following:

– additives added as intended, e.g. preservatives, colours, emulsifiers;
– substances naturally found in food, e.g. amygdalin in almonds, solanine in potatoes, toxins in cap mushrooms;

- substances present in food as a result of human activity, e.g. fertilisers, pesticides, nitrates, veterinary drugs;
- technological agents, e.g. washing and disinfecting agents;
- heavy metals, toxins [van Asselt *et al.*, 2022].
  Physical hazards are all elements that got into food in an accidental unintended way. Their presence may cause physical damage to the organism, e.g. injuries of the oesophagus and mouth. The main sources of physical pollutants are the following:
- raw materials: stones, skins, bones, fish bones, fruit stones, shells, gravel;
- production processes: metal, glass and plastic fragments, screws, splinters, tool parts;
- a human source: jewellery, hairs, nails, cigarette ends;
- the sanitary condition of a building: plaster, paint, glass from lighting lamps, pests, laboratory glass [Gizaw 2019; Sitarz & Janczar-Smuga 2012].

Deliveries of fresh food that can be carried out throughout the entire year are the result of improvement in food storage and transport conditions. An advantage of this development is the elimination of the need of home food preservation and processing for reasons other than the availability of specific foods in local supermarkets. In many cases, plant food is consumed fresh and therefore we should take into account the need to conduct physico-chemical and microbiological tests of a given plant material. In spite of storage techniques applied, spoilage of a food product may occur at each stage of the entire food production chain. These transformations take place as a result of natural biochemical and physical processes as well as under the influence of microbial activity. Therefore, the basis for food spoilage prevention is proper food storage, which is possible only when the properties of raw materials and finished products are known. Modern technologies predominantly incorporate storage at reduced temperature that is continuously controlled. Violation of required thermal regimes and emergence of cyclic and variable temperature effects of the environment during storage and transportation lead to a sharp acceleration of the processes of bacterial spoilage in products, which negatively affects their quality indicators and safety in a sales process [Avilova *et al.*, 2019; Singh *et al.*, 2019]. Storage of plant-based raw materials aims to avoid deterioration of their quality and to ensure their safety by determining the most appropriate temperature and relative humidity as well as by creating specific storage conditions. This can be achieved by reducing the water content of the product to safe levels, cooling or use of modified atmosphere. The plant harvest method and the immediate postharvest handling of a raw material, which determines the level of bioactive substances, have a direct impact on storage efficiency [Lisboa 2018].

Root crops such as carrot, parsley, celery or potato, which are exposed to rotting processes and pests due to their high water content, are among the most frequently stored agricultural commodities. An important element of storage is to provide optimal conditions that will allow life functions in these plants to be stopped. In the case of potatoes, the storage period is the longest period in the entire technological process and hence they are exposed the longest to risks involving qualitative changes. These tubers should be handled with special caution during their transfer because the greatest mechanical damage may occur during this operation, which will have a direct impact on the development of diseases that may substantially reduce their suitability for processing. To protect harvested yields against undesired changes and diseases, which include soft rot, an appropriate microclimate should be created at the place of storage and physical damage to tubers should be prevented [Cwalina-Ambroziak 2022; Gizaw 2019]. As shown in the studies conducted by Wrona [2012] and Kwiatkowski *et al.*

[2015], in the case of carrot the moisture content of a raw material is increasing with increasing storage time. A decrease in the percentage of dry matter was also observed, which results from metabolic processes, particularly cellular respiration. As a result of these reactions, chemical compounds making up storage reserves in plants are lost [Cwalina-Ambroziak 2022].

Plant oils are the most susceptible to qualitative changes caused by storage since particular oil components are chemically very unstable and undergo numerous trans-formations. This happens due to improper temperature, unrestricted light, exposure to oxygen and also a too long storage period. These are factors that reduce the sensory value of oils and lead, among others, to photosensitised oxidation and polymerisation. Products of reactions occurring in oils as a result of improper storage may sometimes lead to the development of tumours, changes in cellular metabolism and nervous system disorders [Chmielewska *et al.*, 2018].

In the case of fruit storage, losses can be both qualitative and quantitative. As far as these raw materials are concerned, the characteristics desired by consumers are fruit firmness, juiciness and tastiness. It is possible to preserve fruits during storage only in the case of long-lasting and high-class fruits. Long-term storage of fruits and vegetables after harvest entails certain consequences associated with the loss of some nutrients. Therefore, the content of phytochemical compounds will largely depend on the cultivation method, but also on factors that occur right after harvest such as expo-sure to light, temperature and humidity [Maghoumi *et al.*, 2022; Esti *et al.*, 2002].

A study conducted by Galani *et al.* [2017] demonstrates that cold storage of pomegranate or sugar beet may significantly reduce the percentage of vitamin C. In some cases, on the other hand, the content of this component may slightly increase and this applies to orange, cabbage and cauliflower. The results were however different for the particular analysis methods and therefore an accurate relationship between the storage method and the level of this vitamin cannot be unambiguously indicated.

A study carried out by Idah *et al.* [2010], in turn, addresses the issue of total ash content, being one of the basic measures of the nutritional value of a product. When analysing the composition of tomatoes and oranges, these authors showed that the ash content gradually decreased with increasing storage period, whereas after two weeks of pot-in-pot storage losses were substantial. The authors of this research also estimated the effects of storage on the protein content in these fruits. The analysis performed on the fourteenth day of storage revealed that the protein content in the samples decreased with increasing storage period. In the case of tomato, losses were about 20%, whereas in the case of orange they could even be 50%. Thus, a decreasing trend is noticeable in the results obtained for the samples tested both in terms of total ash and total protein.

Many meta-analyses reveal that organic farming products are characterised by longer storage life than products originating from conventional farming. This is due to the high metabolic integrity and a better intercellular structure making up the tissues of organic plant raw materials. A higher percentage content of dry matter in organic products is a parameter that confirms these scientific findings [Tal 2018].

Chapter 5

Quality of selected raw materials and food products derived from organic and conventional farms – results of own studies and literature review

5.1 INTRODUCTION

In the chapter, the analysis of the parameters that prove the nutritional value of selected plant products obtained using organic and conventional methods was done. Additionally, the impact of the applied cultivation system on the features of selected fruits related to their storage was examined.

It was hypothesized that ecological farming practises of producing plant-derived food contribute to its higher nutritional value compared to the products made on the basis of raw materials obtained from conventional agriculture systems, and contribute to increasing the share of biologically active substances, which determine the health-promoting properties of the food. This hypothesis was verified, among others, on the basis of the results of qualitative laboratory tests, which covered a number of parameters determining the potential advantages of ecological products. In addition, a storage test of selected "ecological" and "conventional" raw materials was carried out under domestic conditions in order to compare their storability.

5.2 COMPARATIVE ANALYSIS OF RAW MATERIALS AND FOOD PRODUCTS DERIVED FROM CONVENTIONAL AND ORGANIC FARMING

Six popular products or raw materials for food production, such as: millet, canned green peas, edible carrot roots, zucchini fruits, apples and bananas were selected for laboratory qualitative research. Each of the tested materials came both from the conventional and organic farming. The examined materials were obtained from randomly selected food outlets in Poland, assuming that each raw material or product ("conventional" and "organic") comes from at least 30 different sources.

Each test sample of "conventional" and "organic" material was analyzed in 3 repetitions. The results of the studies were compiled separately for each type of material, and the average value and the standard deviation (SD) of the results was calculated. The parameters significant for determining the nutritional value of raw materials or plant-derived food products were assessed according to standard analytical methods (particular methods are described below the tables in which the results are presented). The results of the study were statistically verified by analysis of variance (ANOVA). Excel 9.0 spreadsheet and statistical package Statistica v.10.0 software for Windows StatSoft were used.

DOI: 10.1201/9781003380771-5

Tables 5.1–5.6 present the quality parameters of the examined plant-based products and food raw materials derived from conventional and organic farming systems.

The data presented in Table 5.1 show that millet the groats obtained from products derived from organic farming was characterized by more favorable parameters than the groats from raw materials obtained from conventional farming. Nevertheless, only in the case of two parameters (contents of protein and o-dihydroxyphenols) the difference which favors the "organic" groats was statistically significant. In the case of other parameters, such as contents of dry matter, dietary fiber and carbohydrates, slightly more favorable properties of "organic" groats were observed, but the differences between the organic and conventional variants were statistically insignificant. It is noteworthy that "organic" millet groats had 2.2 percentage point higher protein content than "conventional" ones, and an even clearer difference (by nearly 30%) in favor of the former was observed in the case of o-dihydroxyphenols content. These ingredients are natural antioxidants with anticancer activity, and provide significant health benefits of "organic" millet groats.

Table 5.1 The results of a comparative analysis of the quality of millet groats produced from raw materials derived from conventional and organic farming systems (average value of parameter ± standard deviation) [Kwiatkowski et al., 2022, unpublished data].

No.	Parameter[*]	Conventional cropping[**]	Organic cropping[**]	HSD[***] (p = 0.05)
I	Total solid (%)	85.7 ± 2.4 a	86.2 ± 2.6 a	n.s.
II	Dietary fiber content (g/100 g product)	17.7 ± 0.9 a	18.0 ± 0.7 a	n.s.
III	Protein content (g/100 g product)	10.3 ± 0.3 b	12.5 ± 0.4 a	1.1
IV	Carbohydrate content (g/100 g product)	69.3 ± 1.1 a	70.1 ± 1.5 a	i.d.
V	o-dihydroxyphenols (sum) content (g/100 g product)	0.037 ± 0.004 b	0.052 ± 0.003 a	0.013

[*] the methods of determination of particular parameters are as follows:
Total solid (I) – according to the weight basis method – based on the results of drying the sample at 105°C, according to PN-90/A-75101/03.
Dietary fiber content (II) – according to PN-A-79011-15:1998.
Protein content (III) – according to Approved Method of the AACC, 9th ed.; American Association of Cereal Chemists: Saint Paul, MN, USA [1995].
Carbohydrate content (IV) – according to Luff-Schroorl method described in PN-90/A-75101/07 [Majkowska-Gadomska et al., 2021].
Sum of o- dihydroxyphenols (V) – spectrophotometrically [Singleton et al., 1999] expressed as caffeic acid.
[**] HSD – Honestly Significant Difference;
[***] the results marked with other letter (a or b) are significantly different (p = 0.05);
i.d. – insignificant difference

The analysis of the content of nutritional components in zucchini, presented in Table 5.2, shows that the fruits obtained from organic farming had better quality when compared to those from conventional farming. Although the contents of dry matter and carbohydrates were similar (differences of the values of these parameters in "organic" and "conventional" fruits were statistically insignificant), with a tendency to slightly higher content in "organic" fruits, the contents of other qualitative parameters were significantly higher in the "organic" variant of the zucchini. Considering the differences of particular parameters in detail, the following were noted: dietary fiber content – higher by about 27%, protein content – higher by about 36%, and vitamin A content – higher by about 11%, in zucchini from an organic plantation compared with the fruits derived from conventional plantation. The quoted values of particular quality parameters of "organic" zucchini are a recommendation for consumers to purchase the vegetables from such plantations.

The laboratory analysis confirmed significantly better quality in terms of nutritional parameters of the "organic" carrot than the "conventional" one (Table 5.3).

The carrots obtained from organic farming system were characterized by 2.7 percentage point higher dry matter content, a 13% higher vitamin A content, a 12% higher dietary

Table 5.2 The results of a comparative analysis of the quality of zucchini fruits derived from conventional and organic farming systems (average value of particular parameter ± standard deviation) [Kwiatkowski et al., 2022, unpublished data].

No.	Parameter	Conventional cropping[**]	Organic cropping[**]	HSD[***] (p = 0.05)
I	Total solid (%)	7.9 ± 0.5 a	8.2 ± 0.6 a	i.d.
II	Dietary fiber content (g/100 g product)	1.1 ± 0.1 b	1.5 ± 0.2 a	0.4
III	Protein content (g/100 g product)	0.9 ± 0.1 b	1.4 ± 0.1 a	0.4
IV	Carbohydrate content (g/100 g product)	2.5 ± 0.2 a	2.7 ± 0.3 a	i.d.
V	Vitamin A content (μg/100 g product)	202 ± 0.4 b	226 ± 0.4 a	19

[*] the methods of determination of particular parameters are as follows:
Parameters I – IV were determined according to the methods given below Table 5.1.
The vitamin A content (V) was determined by HPLC-DAD, adapting the HPLC method with UV detection for the determination of niacin in supplements [Chotyakul et al., 2014].
Other notes as given below Table 5.1.

Table 5.3 The results of a comparative analysis of the quality of carrot roots derived from conventional and organic farming systems (average value of particular parameter ± standard deviation) [Kwiatkowski et al., 2022, unpublished data].

No.	Parameter	Conventional cropping[**]	Organic cropping[**]	HSD[***] (p = 0.05)
I	Total solid (%)	10.8 ± 0.5 b	12.5 ± 0.4 a	1.4
II	Dietary fiber content (g/100 g product)	32.0 ± 0.9 b	36.2 ± 0.7 a	4.1
III	Carotenoids content (mg/100 g product)	104.2 ± 8.4 b	120.3 ± 9.3 a	16.1
IV	Vitamin C content (mg/100 g product)	12.9 ± 0.4 b	13.7 ± 0.5 a	0.6
V	Carbohydrate content (g/100 g product)	52.3 ± 1.5 b	57.3 ± 1.3 a	4.2
VI	Phenolic compounds content (mg/100 g product)	814.3 ± 9.4 b	891.2 ± 9.9 a	67.2

[*] the methods of determination of particular parameters are as follows:
Total solid (I), dietary fiber content (II), carbohydrate content (V) – according to the methods given below Table 5.1.
Carotenoids content (III) – spectrophotometrically (UV-120IV spectrophotometer), at wavelength 450 nm [Biehler et al., 2010].
Vitamin C content L-ascorbic acid (IV) – according to the Tillmans method modified by Pijanowski described in PN-90/A-75101/11 [Marku-szewski & Kopytowski 2013].
according to the method given below Table 5.2.
Phenolic compounds content (VI) – by pharmacopoeial method, according to Arnov [Farmakopea Polska VI 2002].
Other notes as given below Table 5.1.

fiber content, a 14% higher carotenoid content, and a 9% higher carbohydrate content. The results obtained confirm the rationality of "organic" carrot cultivation, because consumers receive a raw material of high nutritional and health-promoting value, especially in terms of the content of carotenoids, vitamin C and phenolic compounds, which is more beneficial than in the case of the carrots derived from conventional plantations.

Apples are one of the most popular fruits in Poland and in the whole world. The trade is dominated by the apples from conventional orchards, but the consumer interest in the apples produced in organic orchards is growing year by year. The reasonableness of such interest of the consumers is confirmed by the results of laboratory study (Table 5.4). The comparison of selected qualitative parameters related to nutritional value shows that the apples from organic farming had greater health benefits for humans compared to those from conventional farming. These fruits contained significantly more dry matter (by 1.2 percentage points), and were richer in dietary fiber and vitamin C than conventional apples, by 21% and 13%, respectively. In addition, "organic" apples contained significantly more potassium and iron than "conventional" ones, by 12%, and up to 24%, respectively.

The results of a laboratory study on bananas showed slight quality differences between organic and conventionally cultivated fruits (Table 5.5). Only in the case of the

Table 5.4 The results of a comparative analysis of the quality of apples derived from conventional and organic farming systems (average value of particular parameter ± standard deviation) [Kwiatkowski et al., 2022, unpublished data].

No.	Parameter	Conventional cropping[**]	Organic cropping[**]	HSD[***] (p = 0.05)
I	Total solid (%)	15.1 ± 0.4 b	16.3 ± 0.4 a	0.9
II	Dietary fiber content (g/100 g product)	2.7 ± 0.3 b	3.4 ± 0.2 a	0.6
III	Vitamin C content (mg/100 g product)	169 ± 11 b	194 ± 10 a	18
IV	Potassium (K) content (mg/100 g product)	243 ± 13 b	276 ± 12 a	22
V	Ferrum (Fe) content (g/100 g product)	2.9 ± 0.2 b	3.8 ± 0.1 a	0.7

[*] the methods of determination of particular parameters are as follows:
Total solid (I), dietary fiber content (II) – according to the methods given below Table 5.1.
Vitamin C content, L-ascorbic acid (III) – according to the methods given below Table 5.3.
Parameter IV – photometrically [Maidana et al., 2009].
Parameter IV – colorimetrically [Kozak et al., 2021].
Other notes as given below Table 5.1.

content of dietary fiber, the significantly higher value of this parameter (by 14%) was proven in the bananas from organic farming. Differences between the content of other examined components of fruits from different types of cropping systems were statistically insignificant. Nevertheless, a trend of increased dry matter, vitamin C, calcium and magnesium content in "organic" bananas compared to the "conventional" ones should be noted. The data compiled in Tables 5.1–5.6 also show that the differences in the quality and health-promoting value of "ecological" and "conventional" fruits were the smallest among all the products and raw materials subjected to a comparative analysis.

The conducted laboratory studies proved the higher nutritional and health value of canned green peas, which were produced from raw materials derived from organic farming, compared to those made from peas harvested from conventional crops (Table 5.6).

For each of the eight tested parameters, its value was found to be significantly higher in the "ecological" variant of the product compared to the "conventional" one.

Table 5.5 The results of a comparative analysis of the quality of bananas derived from conventional and organic farming systems (average value of particular parameter ± standard deviation) [Kwiatkowski *et al.*, 2022, unpublished data].

No.	Parameter	Conventional cropping[**]	Organic cropping[**]	HSD[***] (p = 0.05)
I	Total solid (%)	16.5 ± 0.4 a	16.8 ± 0.5 a	n.s.
II	Dietary fiber content (g/100 g product)	3.3 ± 0.2 a	3.8 ± 0.2 b	0.5
III	Vitamin C content (mg/100 g product)	8.6 ± 0.5 a	8.8 ± 0.3 a	n.s.
IV	Calcium (Ca) content (%)	5.0 ± 0.1 a	5.2 ± 0.2 a	n.s.
V	Magnesium (Mg) content (%)	26.7 ± 0.7 a	27.1 ± 0.8 a	n.s.

[*] the methods of determination of particular parameters are as follows:
Total solid (I), dietary fiber content (II), carbohydrate content (V) – according to the methods given below Table 5.1.
Vitamin C content, L-ascorbic acid (III) – according to the methods given below Table 5.3.
Ca content (IV) – photometrically [Maidana *et al.*, 2009].
Mg content (V) – using atomic absorption spectrometer (AAS) [Sbahi *et al.*, 2020].
Other notes as given below Table 5.1.

Table 5.6 The results of a comparative analysis of the quality of canned green peas produced from raw material derived from conventional and organic farming systems (average value of particular parameter ± standard deviation) [Kwiatkowski *et al.*, 2022, unpublished data].

No.	Parameter	Conventional cropping[**]	Organic cropping[**]	HSD[***] (p = 0.05)
I	Total solid (%)	16.2 ± 0.4 b	23.4 ± 0.6 a	5.1
II	Dietary fiber content (g/100 g product)	5.4 ± 0.2 b	6.2 ± 0.3 a	0.5
III	Protein content (g/100 g product)	5.2 ± 0.4 b	6.7 ± 0.3 a	0.6
IV	Carbohydrate content (g/100 g product)	12.3 ± 0.5 b	17.1 ± 0.4 a	4.2
V	Vitamin C content (mg/100 g product)	31.0 ± 0.6 b	34.4 ± 0.7 a	3.2
VI	Vitamin A content (μg/100 g product)	62.3 ± 0.4 b	68.1 ± 0.4 a	5.6
VII	Vitamin B3 content (mg/100 g product)	2.1 ± 0.1 b	2.7 ± 0.2 a	0.5
VIII	Magnesium (Mg) content (mg/100 g product)	24.4 ± 0.4 b	29.2 ± 0.3 a	3.9

[*] the methods of determination of particular parameters are as follows:
Total solid (I) and dietary fiber content (II), protein content (III) and carbohydrate content (IV) -according to the methods given below Table 5.1.
Vitamin C (V) content – according to the methods given below Table 5.3.
Vitamin A (VI) content – according to the methods given below Table 5.2.
Vitamin B3 (niacin) – adapting HPLC with UV detection method for supplements [Çatak 2019].
Mg content (V) – according to the methods given below Table 5.5.
Other notes as given below Table 5.1.

"Organic" peas had a higher dry matter by as much as 7.2 percentage points, contained 13% more dietary fiber, 22% more protein and 28% more carbohydrates, compared to the "conventional" variant. In addition, it was significantly richer, than "conventional" peas in vitamins C, A and B3, as well as magnesium, by: 10%, 9%, 23% and 17%, respectively.

Both the own comparative research, and studies conducted by other authors confirm the higher content of vitamin C in the case of "organic" foods. In the own analysis, this parameter was: 13.7 ± 0.5 mg/100 g product for "organic" carrot roots and 12.9 ± 0.4 mg/ 100 g product for their "conventional" variants, 194 ± 10 mg/100 g product for the apples obtained from organic orchards and 169 ± 11 mg/100 g product from conventional orchards, 34.4 ± 0.7 mg/100 g product for canned peas, which were produced from raw material from organic farming and 31.0 ± 0.6 mg/100 g product for the equivalent of this product made from the plants from conventional crops. In authors' own research, no significant differences in the content of vitamin C were observed only in the case of bananas. Adamczyk & Rembiałkowska [2005] also showed a higher content of vitamin C in the apples from organic farming. The content of this vitamin in the "organic" variant of the fruits examined by them was 22% higher than in the fruits from conventional cropping systems. In addition, a higher share of dry matter, amounting to 15.13%, was found in "organic" apples, compared to "conventional" fruits, in which this parameter was equal to 13.95%. A higher content of L-ascorbic acid for crops from organic farming has also been found in studies on other raw materials. Kolbe & Stephan-beckmann [1997], and Rodrigues *et al.* [2016] showed that the potatoes grown in accordance with the principles of organic farming had a higher percentage of vitamin C. As stated in the research of Bobrowska-Korczak *et al.* [2016], up to nine out of the seventeen food products produced organically were characterized by a higher content of this ingredient compared to raw materials from traditional cropping system. In the relevant literature, there are still few papers dealing with the relationship between the content of vitamins and the method of cultivation, but the available research results so far confirm the beneficial effect of organic farming in this respect.

Another analyzed parameter was the content of dry matter. In this research, significant differences between raw materials or products of "conventional" and "organic" variants were obtained for carrot roots (variants from organic farming: 12.5 ± 0.4%, from conventional farming: 10.8 ± 0.5%), apples (variants from organic farming: 16.3 ± 0.4%, from conventional farming: 15.1 ± 0.4%) and canned peas (raw material from organic farming: 23.4 ± 0.6%, from conventional farming: 16.2 ± 0.4%). An analysis of this parameter, both in "organic" and "conventional" food is often used by researchers to asses the influence of cultivation system on quality of raw materials or products. For example, in the research of Rochalska *et al.* [2011], no significant differences were observed between the content of dry matter in strawberries from both types of cultivation systems. In the present research, no significant differences were found for millet groats, zucchini and bananas. As stated by Prędka & Gronowska-Senger [2009], the dry matter content was higher in "organic" products, but the difference was not statistically significant. Only in the case of boiled carrots, a significant difference can be noticed – the carrots from organic farming subjected to heat treatment contained 12.18 ± 0.32% of dry matter, and the carrots from traditional farming – 11.23 ± 0.37%. Masaba & Nguyen [2008] did not show differences in the value of this parameter in the roots of "organic" and "conventional" carrots. However, according to Kazimierczak *et al.* [2017], the analyzed material in the form of herbal spices had significantly higher dry matter content; thus it had lower water content, which contributes to the overall quality of food products – both in terms of storage and the influence of consumer's health. However, as research shows, the content of dry matter in the products and raw materials obtained from organic and conventional farming is often at a comparable level, and depends not only on the cultivation methods, but also on the plant variety, as well as soil parameters and climatic conditions.

The content of carbohydrates directly affects the taste of the analyzed products. In this research, it was noticed that in the case of carrots and canned peas, the value of this parameter was clearly higher for the variants obtained using organic cropping methods. Similar conclusions on the influence of the type of cropping system on the content of carbohydrates in crops come from the study of Hallmann & Rembiałkowska [2007], conducted on red onion. The authors found that the value of this parameter in "organic" onion was 47% higher than in "conventional" one.

In authors' research, the content of protein in millet groats and canned peas was also analyzed. In both cases, the percentage value of this parameter was significantly higher in "organic" products. This parameter depends mainly on soil fertilization, i.e. on the amount of nitrogen supplied and absorbed by the plant root system. In some cases, according to own studies and other authors' research, the protein content in the crops from organic farming exceeds those measured in the raw materials from conventional farming. However, not all researchers had the same observations. Murawska *et al.* [2015] found that a higher percentage of protein (by 17.4%) was found only in the roots of a carrot from an organic farming system than in the traditional one, while in other tested raw materials, a higher value of this parameter was not found in products obtained from organic cropping systems. According to Remiałkowska *et al.* [2012], this may be due to the fact that the plants cultivated with traditional methods, due to high nitrogen availability in soil, and as a consequence, increased uptake of this nutrient from the soil, are capable of many metabolic transformations, as a result of which the content of proteins in growing. However, the authors claim that the protein in the case of the "organic" crops is of much better quality due to the favorable amino acid composition.

In the review of the literature, organic food is defined as much richer in biologically active compounds which contribute to good functioning of the body. The results of our research confirm these reports. The "organic" millet groats tested by the authors were characterized by a much higher amount of o-dihydroxyphenols, while the carrot roots were richer in total phenolic compounds. According to Benbrook *et al.* [2008], these components play an important role in the prevention of cancer, and ecological farming practices increase the content of these substances in agricultural products. The conducted laboratory analysis showed that content of total phenolic compounds in the carrot roots from conventional cropping was 814.3 ± 9.4 mg/100 g, and in the carrot roots from organic farming was 891.2 ± 9.9 mg/100 g. Similar relationships were shown by Faller & Fialho [2010], where the amount of these compounds was much higher in raw materials such as: carrots, onions, tomatoes and cabbage from organic farming. The higher content of phenolic acids in the roots of "organic" carrots is also confirmed by Sikora *et al.* [2009]. In addition, in the study material, the authors showed a significantly higher content of carotenoids in the vegetables from organic farming. These observations were also confirmed in this research.

Carotenoids were also the compounds which concentration was significantly higher in the carrot roots derived from organic cropping systems than in "conventional" variants. Carotenoids play the role of natural antioxidants; hence they are a desirable food ingredient, the presence of which is found in the largest amounts in carrot roots. Their higher content was also found by Kazimierczak *et al.* [2018], who analyzed the composition of carrot juices produced from raw materials of various origins.

The food produced in organic farming systems is usually characterized by a higher content of mineral compounds in relation to food obtained in conventional systems. In the conducted research, a higher share of potassium and iron in "organic" apples (by 12% and as much as 24%, respectively) and magnesium in "organic" canned peas (by 17%) was shown. The "organic" tomatoes and carrots examined by Harecia *et al.* [2011], also contained significantly more phosphorus and potassium, whereas Masaba & Nguyen [2008] found higher amounts of potassium and calcium in "organic" cabbage, carrots and lettuce.

5.3 INFLUENCE OF STORAGE ON THE QUALITY OF SELECTED PRODUCTS FROM CONVENTIONAL AND ORGANIC FARMING

For the storage tests two types of the food products which were examined in the laboratory study – apples and bananas – were selected. Both the apples and bananas came from two farming systems – organic and conventional. The storage test of bananas was carried out under home conditions, in the period 04-15.04.2022, while the storage test of apples was performed in the period 04-28/04/2022. The research began on the day of fruit purchase. The fruit was stored at temperature range from 16 to 18°C. The study consisted in observing the changes in fruit appearance, as shown in the photographs (Photos 5.1–5.16) taken with a SONY DSC-H300 camera. The frequency of taking photographs depended on the type of fruit and the speed of changes in its appearance.

Bananas are fruits with a stronger relationship between storage time and changes in their appearance. The changes appearing along with the time of the experiment are presented in the photographs (Photos 5.1–5.11).

Photo 5.1 Bananas from a conventional (left) and organic (right) plantation on the day of their purchase (Source: own archive).

Photo 5.2 Bananas from a conventional (left) and organic (right) plantation on the second day after their purchase (Source: own archive).

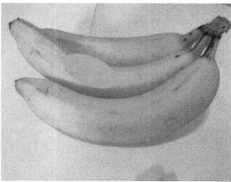

Photo 5.3 Bananas from a conventional (left) and organic (right) plantation on the third day after their purchase (Source: own archive).

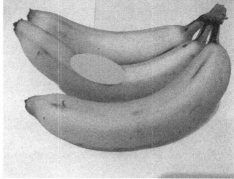

Photo 5.4 Bananas from a conventional (left) and organic (right) plantation on the fourth day after their purchase (Source: own archive).

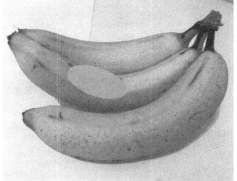

Photo 5.5 Bananas from a conventional (left) and organic (right) plantation on the fifth day after their purchase (Source: own archive).

Photo 5.6 Bananas from a conventional (left) and organic (right) plantation on the sixth day after their purchase (Source: own archive).

Photo 5.7 Bananas from a conventional (left) and organic (right) plantation on the seventh day after their purchase (Source: own archive).

Photo 5.8 Bananas from a conventional (left) and organic (right) plantation on the eight day after their purchase (Source: own archive).

Photo 5.9 Bananas from a conventional (left) and organic (right) plantation on the ninth day after their purchase (Source: own archive).

Photo 5.10 Bananas from a conventional (left) and organic (right) plantation on the tenth day after their purchase (Source: own archive).

Photo 5.11 Bananas from a conventional (left) and organic (right) plantation on the twelfth day after their purchase (Source: own archive).

As it can be observed in the photographs (Photos 5.1–5.11), the rate of morphological changes which appeared in bananas did not differ significantly in the case of fruits from conventional and organic farming. Only in the last days of the storage test, a more intense change in the color of the "conventional" fruit could be observed, which are the points in favor of the "organic" fruit, and may indicate its longer shelf life.

The changes occurring in the apples subjected to the storage test, related to the appearance of the fruit, are shown in Photos 5.12–5.16

Photo 5.12 Apples from a conventional (left) and organic (right) plantation on the day of their purchase (Source: own archive).

Photo 5.13 Apples from a conventional (left) and organic (right) plantation on the eight day after their purchase (Source: own archive).

Photo 5.14 Apples from a conventional (left) and organic (right) plantation on the sixteenth day after their purchase (Source: own archive).

Photo 5.15 Apples from a conventional (left) and organic (right) plantation on the twentieth day after their purchase (Source: own archive).

Photo 5.16 Apples from a conventional (left) and organic (right) plantation on the twenty-fifth day after their purchase (Source: own archive).

Apples, similarly to bananas, are raw materials in which no clear changes have been observed that could indicate different storage durability caused by applied farming system. Fruit from both organic and traditional orchards were stored in a similar way during the study, and no clear physical changes, such as fruit color or degree of skin wrinkling, which might favor organic or conventional raw material, were found.

Own research on changes in the storage of raw materials of organic and conventional origin was not fully consistent with the results of laboratory tests performed on the same raw materials, which are discussed in subchapter 5.2. According to the current state of knowledge, the dry matter content affects the storability. Therefore, "organic" fruits, having a higher percentage of dry matter, should have a higher shelf life. The "organic" apples used in the study had significantly more dry matter (by 1.2 percentage points) than the apples obtained by conventional methods. As shown in the photographs (Photos 5.12–5.16), it did not affect the morphological changes of the fruit. The qualitative tests showed no significant differences in the content of dry matter in bananas, which translated into a similar storage life of this fruit produced by conventional and organic methods.

Chapter 6

Quality of honey and other bee products derived from conventional and organic beekeeping farms

6.1 INTRODUCTION

Honey is commonly considered to be a valuable food product that not only has nutritional and taste values, but above all has a positive impact on health, strengthens the body, exhibits antibacterial effects, accelerates healing of wounds, lowers blood pressure, aids digestion etc. Multiple beneficial health effects are also attributed to other bee products, such as flower pollen, propolis or bee bread. This is determined by the specific components contained in these products, whose composition and concentration predominantly depend on the type of raw material processed by bees. The quality of bee products is also affected by conditions in a beekeeping farm, including apiary management, the approach to control of bee diseases, the quality of the environment surrounding an apiary as well as product harvesting and storage conditions. The different approaches to these aspects discriminate beekeeping farms into conventional and organic ones. The requirements with regard to organic apiaries in the European Union are set out in the Regulation (EU) 2018/848 of the European Parliament and of the Council of 30 May 2018 on organic production and labelling of organic products and repealing Council Regulation (EC) No 834/2007 (OJ L 150/2). Organic beekeeping farms must meet a number of requirements. In beekeeping, preference is given to the species *Apis mellifera* and its local ecotypes. It is important that an apiary should be located in such a way that within a radius of 3 km from the apiary site, nectar and pollen sources consist of organically produced crops, spontaneous vegetation, forests or crops treated with low environmental impact methods. No sources that may lead to the contamination of bee products or to the poor health of the bees can be located near the apiary site. The hives and materials used in beekeeping should be made basically of natural materials presenting no risk of contamination to the environment or the bee products, while the beeswax for new foundations should come from organic production units. Synthetic chemical repellents may not be used during honey extraction operations. Appropriate conditions should be created for bees which allow them to survive the winter, leaving sufficient reserves of honey and pollen at the end of the production season, while feeding of bee colonies is only allowed where the survival of the colony is endangered due to climatic conditions and in such case only organic products may be used. Bee products may be sold as organic ones only after one year from conversion to organic beekeeping.

This chapter presents the characteristics of the composition and properties of honeys and other bee products essential for their quality, in particular their

DOI: 10.1201/9781003380771-6

health-enhancing values. A comparative analysis has also been made based on the results of the authors' study of the quality of two honey varieties, acacia honey and lime honey, produced using organic and conventional methods. The chapter ends with the presentation of the results of surveys conducted on a group of several hundred respondents which related to the presence of honey and other bee products in diet, factors affecting the purchase decision, and ways of using honey in a household.

6.2 NUTRITIONAL AND HEALTH-ENHANCING PROPERTIES OF HONEY AND BEE PRODUCTS

6.2.1 *The composition and properties of honey*

Honey is a sweet product of natural origin, produced by bees of the species *Apis mellifera*. It is made by combining bees' own specific substances and flower nectar, or excretions of plant-sap-sucking insects on living parts of plants, or secretions of living parts of plants, which are stored and dehydrated, and ripen in honeycombs. According to the quality standards applicable in Poland (Regulation on detailed requirements for commercial quality of honey, Journal of Laws of 2003 No. 181, items 1772 and 1773, as amended; Journal of Laws of 2015, item 850), which comply with the guidelines of the Council Directive 2001/110/EC of 20 December 2001 (OJ L10/47, as amended), 2 types of honey can be distinguished depending on its origin: blossom, or nectar honey, and honeydew honey. Moreover, honey varieties are distinguished depending on the plant from which honey was produced: polyfloral, buckwheat, lime, acacia, rape or heather honey, or depending on the plant from which honeydew was collected, e.g. coniferous honeydew honey or deciduous honeydew honey. The honey type or variety is determined by the following: taste, smell, colour and consistency. The variety of nectar honey depends on which pollen has the highest percentage in the honey. Marketed honey must meet the organoleptic and physio-chemical requirements (Table 6.1) set out in the above-mentioned Regulation ... [2015, item 850]

The diastase number determines the presence of the diastase enzyme in honey, which indicates the freshness of the product, and it is of essential importance for health-enhancing properties. Low diastase activity may mean the development of microorganisms in honey, which may result in fermentation. This indicator determines the amount of starch that is hydrolysed under the influence of diastatic enzymes contained in 1 g of honey [Tosi *et al.*, 2008].

Different botanical varieties of honey are present in the market worldwide. According to Consonni *et al.* [2019], about 300 varieties of honey have been identified taking into account pollen composition. Two types of beekeeping practices, which generally differ in terms of approaches to using chemicals, are available in Europe: conventional and organic ones. In the organic beekeeping practice, only natural products are allowed to be used. For example, natural phytochemicals such as eucalyptol or thymol can be applied against the Varroa destructor, clean wax paper without chemical contamination must be used for the deposition of honey within the beehive, and only pollen or honey could be used for feeding the bees as indicated in the relevant EU regulation [Regulation (EU) 2018/848]. Additionally, the environment around the apiary must be free of chemical contamination.

Table 6.1 The physico-chemical requirements for honey [Regulation on detailed requirements ... , Journal of Laws of 2015, item 850].

Specification	Criteria
Moisture	not more than 20%, but not more than:
	1) 23% – in heather honey and in baker's honey
	2) 25% – in baker's heather honey
Glucose and Fructose Content (sum of both)	not less than: 60g/100g – in nectar honey, and 45g/100g – in honeydew honey and honeydew-nectar honey
Sucrose Content	not more than 5g/100g, but not more than:
	1) 10g/100g – in honey originating from: black locusts (*Robienia pseudoacacia*), alfalfa (*Medicago sativa*), Menzies Banksia (*Banksja menziesii*), sweetvetch (*Hedysarum*), river red gum (*Eucalyptus camaldulensis*), leatherwood (*Eucryphia lucida, Eucryphia miliganii*), *Citrus* spp.;
	2) 15g/100g – in honey originating from: lavender (*Lavandula* spp.), borage (*Borago officinalis*)
Water Insoluble Solids Content	not more than:
	g/100g, but not more than:
	0.5g/100g – in pressed honey
Electrical Conductivity	not more than 0.8 mS/cm, except for honeys and their blends mentioned below, but not less than 0.8 mS/cm – in honeydew honey, chestnut honey and their blends with other honey varieties. Electrical conductivity is not determined for honey originating from strawberry tree (*Arbutus unedo*), heath (*Erica*), eucalyptus, lime (*Tilia* spp.), heather (*Calluna vulgaris*), leptospermum, tea tree (*Melaleuca* spp.)
Free Acids	not more than 50 mval/kg, but not more than 80 mval/kg – in baker's honey (industrial honey)
Diastase Number (according to Schade scale)	not less than 8, except for baker's honey (industrial honey), but not less than 3 in honey with naturally low enzymatic activity and an HMF content of not more than 15 mg/kg
Content of 5-Hydroxy Methyl Furfural (HMF)	not more than 40 mg/kg, except for baker's honey (industrial honey), but not more than 80 mg/kg – in honey originating from tropical climate regions and in blends of such honeys

It is estimated that honey has been used not only as food but also for therapeutic purposes for at least 4000 years. Nowadays, it is often used in home methods to treat colds as well as in exhaustion and fatigue. It contains a substantial amount of simple sugars and its calorific value is 324 kcal/100g. Honey is frequently used to aid in the treatment of diseases of the digestive, respiratory, circulatory and urinary systems. It is known to have antibacterial properties and shows the best effects against Gram-positive bacteria, e.g. *Staphylococcus aureus* and *Streptococcus pyogenes*. Gram-positive bacteria exhibit greater resistance to its action. Honey acts against *Bacillus anthracis*, *Trichomonas vaginalis* and also *Mycobacterium tuberculosis*. Honey's antibiotic properties are attributed to its physico-chemical and chemical properties and also to the substances coming from honeydew from coniferous trees and essential oil-producing plants – thymol, camphene, menthol and pinene. Its antibiotic activity is significantly enhanced after honey is diluted with water since the strength of this activity increases from 6 to even 220 times. Dark honey shows much better antioxidant activity compared to light honeys. To preserve the health and nutritional values of honey, it is recommended to be consumed raw. The maximum temperature at which honey does not lose the above-mentioned properties is lower than or equal to 40°C [Can *et al.*, 2015; Hołderna-Kędzia & Kędzia 2006; Vasić *et al.*, 2019].

6.2.2 The characteristics of other bee products

Flower pollen is produced by flowers in large amounts in anthers and pollen grains are male reproductive cells of seed-producing plants. When collecting pollen, bees moisten the pollen with honey or nectar and form it into lumps called pollen loads. After the pollen has been transported to the beehive, it is then compacted with the addition of honey in the honeycomb cells. Following lactic acid fermentation, the pollen is transformed into bee bread [Harasim & Kwiatkowski 2020]. Bees gather most pollen in the morning and to form a pollen load of appropriate size, one insect must visit from 7 to 120 flowers. It is easier for bees to cooperate with entomophilous flowers whose pollen is larger and covered with fats and special balm. The formation of pollen loads from anemophilous plants takes longer and this can be done only on windless days [Komosinska-Vassev *et al.*, 2015]. This product is an essential and necessary ingredient of bee food and its components contribute to performing different life functions and feeding larvae. Within a year, one bee colony can use even 30–50 kg of pollen, whereas humans are able to harvest only 1–3 kg. Pollen load colour depends on the plant from which the pollen was gathered and whether it was gathered from open anthers or anthers bitten open by the bee. The most popular pollen colours are yellow and orange, though greenish pollen originating from pear trees, crème-grey from raspberry, dark grey or black from poppy, red-orange from mignonette, dark red from chestnut, and dark blue from viper's bugloss can also be found [Paradowska *et al.*, 2014]. The composition and properties of pollen depend on the type of plant from which it was collected as well as on the soil, climate and pollen collection time [Kocot *et al.*, 2018; Wang and Li 2011].

Flower pollen consists of:

- glucose (27%)
- proteins (23%)
- water (18%)
- fibre (18%)
- mineral salts (18%)
- fats (5%)
- other substances (4%) [Komosinska-Vassev *et al.*, 2015].

Pollen is a product rich in magnesium, potassium, calcium, sodium, vitamins A, C, E, PP, B1, B2, B6 and H, and folic acid. It cannot replace sustainable nutrition, but it is recommended as a diet supplement because pollen exhibits nutritional, antibacterial, anti-inflammatory, regenerative and anti-cancer effects. Medical studies have revealed that pollen aids in the treatment of anaemia, stomach ulcers, cardiovascular diseases, mental and sexual performance, liver diseases, duodenal ulcers, prostate inflammations as well as nervous, mental, allergic and respiratory diseases [Bakour *et al.*, 2022; Khalifa *et al.*, 2021].

Royal jelly is a secretion of the pharyngeal glands of bees and it is produced by worker bees. This substance is secreted between the 7th and 14th day after a worker bee eats through the cell. Drone and worker bee larvae are fed with royal jelly during the first 3 days of their life, while the queen is fed during the larval stage and the reproduction period. In apiaries that specialise in the production of royal jelly, a large number of queen cells are built and then royal jelly is harvested from them on the third day from placing the queen larva in the cell. At that time, royal jelly is of highest quality and there is the largest amount of it. After royal jelly has been harvested from the queen cells, it should be immediately transferred to tightly closed containers and stored in a refrigerator for 7 days [Harasim & Kwiatkowski 2020].

The chemical composition of royal jelly for specific larvae is different. It is a substance with the consistency of sour cream that may have different colours, e.g. blue-white, dirty grey, light crème or light yellow. Royal jelly contains 67% of water and its dry matter consists of the following: 17–45% proteins, 3.5–19% fats and 18–52% carbohydrates. It contains more than 20 amino acids, including all exogenous ones, vitamins and biologically active substances: proteinases, glucose oxidase, catalase, amylase and acid phosphatase. Royal jelly also has small amounts of sex hormones: testosterone, estradiol and progesterone. Research has demonstrated that there are 22 elements in royal jelly and that it is the richest in potassium. The other macro- and micronutrients making up its composition include the following elements: sodium, phosphorus, sulphur, calcium, iron, zinc, manganese and copper [Pasupuleti *et al.*, 2017].

This product has a valuable impact on human health. It provides nutrients and is readily absorbable by the body. Evan small doses, taken regularly, increase the amount and volume of red cells as well as the level of haemoglobin and iron in blood. Royal jelly has an excellent effect on the circulatory system, shows antiatherogenic activity, and lowers the cholesterol and sugar level. It aids metabolism and the immunological system, regulates the level of sex hormones, and increases sperm production. Royal jelly performs well in tissue regeneration, accelerating healing of wounds, while fractures may heal up to 3 times faster, and contributes to collagen restoration. It has antibacterial properties against *Staphylococcus aureus*, *Streptococcus pneumonaide*, *Bacillus anthracis*, *Escherichia coli*, Salmonella sp. and Proteus sp. This bee product inhibits the growth of moulds and yeasts and acts against the mumps, herpes and influenza viruses [Ahmad *et al.*, 2020; Guo *et al.*, 2021]. It owes its bactericidal and bacteriostatic properties to the presence of 10-hydroxydecanoic acid, hydrogen peroxide and peptides with antibacterial properties: jelleine and royalisin [Fratini *et al.*, 2016a; Gliński *et al.*, 2011]. Royal jelly also finds its application in cosmetology as an additive to moisture creams. A study conducted by Bocho-Janiszewska *et al.* [2013] demonstrates that creams prepared with the addition of lyophilised royal jelly exhibited beneficial application properties. This product contributes to skin condition improvement, cooperates with other ingredients of emulsion, and can be successfully used as an additive to creams.

Beeswax is a lipid secreted by young 12 to 18-day-old wax-producing bees. It is a liquid mixture of fatty acids, hydroacids, higher alcohols, hydrocarbons and esters. Four pairs of wax glands are located on the underside of their abdomen. Wax is secreted there and it solidifies on chitin platelets. Following the beeswax solidification, bees remove wax scales and they chew the wax material using their mandibles before they start comb construction. During a season, a bee colony can produce 2–3 kg of wax on average, but under Polish conditions only about 0.5 kg is achievable [Cornara *et al.*, 2017]. This product comes in different colours, from almost white through yellow to black. In commerce, two types can be found most frequently: general application wax and wax of cosmetic and pharmaceutical grade. It is characterised by different properties depending on from what region of the world it comes [Fratini *et al.*, 2016b].

The history of beeswax dates back to antiquity when it was used for mummification, construction of boats and making candles or a paint binder.

In mediaeval Europe, it performed the role of a unit of account in trade and payment of taxes. In unconventional medicine, positive effects are achieved by chewing a honeycomb or wax cappings, which aids in glossitis and tonsillitis, gum and teeth strengthening as well as in mucositis and pharyngitis. Another example is an ointment prepared from wax, Rivanol and plant oil, which aids in the treatment of difficult healing wounds, frostbites and ulcerations. Osteoarthrosis and musculoskeletal diseases can be alleviated by beeswax and propolis compresses at a ratio of 10:1. Wax candles give off a pleasant aroma and are recommended in upper respiratory tract inflammations and for asthma sufferers. Among all bee products, wax is the least allergenic [Dumitru *et al.*, 2022]. It is also used in the cosmetics industry as an additive to cosmetics that performs the function of a regulator of consistency and viscosity. Apart from cosmetics, beeswax also finds application in the pharmaceutical industry as an ingredient of pomades, plasters and ointments. It has protective and lubricating properties [Dumitru *et al.*, 2022; Svečnjak *et al.*, 2019].

Propolis – also known as bee glue – is a resinous and balsamic substance collected by bees from young shoots and buds of willow, alder, oak, ash, poplar and birch. Forager bees transport lumps of resin in the same way as pollen loads, then they mix the resin with secretions from their salivary glands and use it to seal the nest. Besides, propolis is also used to mummify dead intruders, thus preventing their decomposition in the beehive when their carcass is too large and heavy to remove it outside the hive, e.g. a mouse. The word "propolis" is of Greek origin and it is composed of "pro" that means "before" and "polis" that means "city", which in this case means the beehive and its defence [Anjum *et al.*, 2019; Pasupuleti *et al.*, 2017].

Propolis occurs in the form of thick and viscous resin with a specific smell and colour that depend on the type of plant from which the lumps of resin were gathered. Depending on the temperature, it has a different structure: below 15°C – hard and brittle, above 36°C – soft and plastic, at a temperature of about 70°C – it is in semiliquid form, while at 90°C – it melts [Hossain 2022].

Propolis primarily consists of:

- resin (40–50%)
- waxes (20–30%)
- polyphenols (14–16%)
- polysaccharides (2–2.5%)
- elements, e.g. iron, selenium, magnesium, copper, phosphorus, silicon, manganese, cobalt and zinc (whose content particularly varies depending on the plant species) [Shruthi and Suma 2012; Szeleszczuk *et al.*, 2013; Wojtacka 2022].

One in vitro study investigated the effect of an ethanolic extract of propolis on the adhesion and growth of *Giardia duodenalis*. It was shown that this preparation inhibited the growth of parasites and promoted their detachment from intestinal epithelium cells. Its efficacy was also confirmed by clinical trials in children and adults where the cure rate was 52 – 60%, while in the case of the drug this rate was 40% [Freitas *et al.*, 2006; Paulino *et al.*, 2015]. Propolis also finds its application in gynaecological treatment.

It has antibiotic, antifungal and pain alleviation properties. Propolis can be used in the treatment of recurrent vulvovaginal candidiasis [Capoci *et al.*, 2015]. Thanks to its antibacterial properties, propolis has the ability to reduce the development of plaque and pathogens causing gingivitis and hence it is used as an additive in toothpastes and mouthwashes [Pasupuleti *et al.*, 2017; Pereira *et al.*, 2011]. It also finds application in dermatology, which is based on its antiallergic, anti-inflammatory, collagen synthesis-stimulating and acne control properties [Ali *et al.*, 2015].

Bee bread has the form of hard lumps with an irregular shape and a dark brown colour, and it is made from flower pollen. Together with the earlier mentioned flower pollen, it is the basic source of protein for bees. Apart from nutrients, it contains other compounds necessary to produce royal jelly, hormones, enzymes and pheromones. During the late autumn period, insects use this food to increase their tissue, which is to ensure their good condition during wintertime. Bee bread helps fight diseases that threaten bees from winter to spring. The remaining reserves can be used to feed larvae and also nurse bees [Bakour *et al.*, 2022].

Flower pollen is transported to the beehive in the form of pollen loads by forager bees where it is stored in empty comb cells and subsequently mixed with secretions from the salivary glands and honey, compacted and secured with a thin layer of honey protecting against air access. The comb cell is filled with such a mixture to ¾ of its depth. If bee bread is to be used as winter reserves, bees fill up the cells with ripe honey and cap them with wax. Under anaerobic conditions, the cell content undergoes lactic acid fermentation and in this way bee pollen is made, called bee bread [Dranca *et al.*, 2020]. Thanks to fermentation, this product becomes more readily absorbable than pollen because chemical reactions occur in the nutrients contained in it. They lead, among others, to destruction of the pollen coat. After several weeks, the protein content in the bee bread is about 15% lower than in the pollen, but the content of amino acids and peptides increases, while lactic acid is formed from simple sugars, reducing the pH of the obtained product from a value of 6.3 to 4.3 [Socha *et al.*, 2018].

Bee bread is a necessary food ingredient for older larvae and young bees, whereas in the case of humans it is recommended in problems with appetite, fatigue, intestinal diseases and anaemia. It exhibits stronger bactericidal and bacteriostatic properties compared to pollen [Mărgăoan *et al.*, 2019]. Bee bread also finds its application in the therapy of proctological and urological diseases. A study conducted on animals shows that bee bread exhibits hypolipidemic activity. Clinical trials on humans confirmed this activity in 20–35% of patients [Urcan *et al.*, 2017]. The patients that did not respond to the fenofibrate drug achieved better trial results after they were given bee bread – a lowered level of cholesterol and lipids by 20–30% and decreased clumping of blood platelets by 30% [Komosinska-Vassev *et al.*, 2015]. Given together with antidepressants, this product allows drug doses to be reduced and improves the overall health condition in a short time. Thanks to these properties, the risk of prescription drug addiction or side effects is reduced. In the case of treatment of a chronic alcohol disease, small doses of bee bread and sedatives result in milder and shorter withdrawal symptoms as well as in replenishment of vitamin and nutrient deficiencies in the body caused by alcoholism [Khalifa *et al.*, 2020; Komosinska-Vassev *et al.*, 2015; Wójcicki 1999].

6.2.3 *Apitherapy*

Bee products have been used in all fields of medicine across the world for 6000 years. At the same time, apitherapy has now gained a scientific basis. Honey and bee products contain substances that have both nutritional values and detoxification, bactericidal, antiviral, antifungal, analgesic, regenerative and preventive properties. Comprehensive use of honey and bee products expands the spectrum of their action and gives very good therapeutic effects. This involves the use of bee products in the treatment and prevention of diseases. Bee products such as, e.g., honey, bee bread, flower pollen, propolis and wax contain many substances beneficial for the human body. Apart from their content of nutrients, honey and bee products are rich in vitamins, minerals, amino acids, proteins, lipids, enzymes, organic acids and antioxidants. Studies indicate that people who support their immunity in a natural way, including the use of apitherapy and supplementation, resort to general preventive treatment much less frequently [Rozman et al., 2022; Soares et al., 2017; Trumbeckaite et al., 2015].

A study aimed to investigate the practice of apitherapy was conducted in Germany. The surveyed respondents were German beekeepers, whereas the investigated products included honey, flower pollen, propolis, royal jelly and bee venom. A questionnaire was published in 3 German beekeeping journals and the readers were asked to complete it. Altogether, 1059 questionnaires were received. This study demonstrates that honey gives the most effective and therapeutic effect, followed by propolis, flower pollen and royal jelly. Beekeepers were also asked for which conditions they would apply propolis and flower pollen. They responded that they used propolis most frequently to treat colds, wounds and burns, sore throats and gum disorders as well as a general prophylactic. As far as flower pollen is concerned, it was also used as a general prophylactic and sometimes to treat prostate diseases. In this study, no adverse effects were reported and the potential benefits arising from application of bee products were supported by the positive experience of a large group of beekeepers who used some of these products to treat a wide range of diseases [Hellner et al., 2007].

Apitherapy also includes respiratory therapy, very popular in Ukraine, Russia and Austria. A person who wants to use this therapy method enters a small house with the size of an allotment shed. Ten bee colonies, e.g., live and work in such a house. The hive entrances are located outside, whereas inside the house the beehives are protected with mesh, and hence a person staying inside the house is not exposed to bee stings. The air in the house is saturated with the hive microclimate. It is a unique form of sauna where one breathes in therapeutic vapours when the bees are processing the nectar gathered in meadows into honey and also a unique microclimate composed of volatile substances from the nectar and propolis [Harasim & Kwiatkowski 2020]. The amount of therapeutic substances depends on the time of the year and the air temperature outside. It is assumed that out of 3 litres of nectar, bees produce 1 litre of honey, that is, 2/3 of the therapeutic mass evaporates. When it is a stand-alone beehive, therapeutic substances "go into the air", but thanks to the location of the hives in such a house, the vapours remain in the chamber. Apart from therapeutic substances, bees may also soothe with their vibrations. The "good noise" of microvibrations is created by the movement of the insects' wings. During the therapy time in the shed, you can lie down or sit. Apitherapy is becoming increasingly popular in some agritourism farms in Poland that provide a beekeeping related offer [Ali & Kunugi 2020; Weis et al., 2022].

Bee venom therapy found its application already 5000 years ago. The idea was based on the conviction that beekeepers hardly suffered from rheumatism or problems with joints. Venom is injected either directly via bee stings or by extracting venom with an electric stimulus and then injecting it with a syringe [Wehbe et al., 2019]. Bee venom consists of a mixture containing more than 18 active components, including proteins, peptides, enzymes, amines, sugars, phospholipids, pheromones and volatile compounds. Apamin and melittin are peptides that are unique components of apitoxin

[Nainu *et al.*, 2021]. Venom, as a complex natural substance containing a substantial amount of proteins, is difficult in processing and requires a lot of attention to preserve as many valuable components as possible. Chemical processing can easily cause loss of proteins and therefore before processing each technological process, venom component and chemical agent as well as their mutual interactions should be analysed to minimise potential losses [Łukasiewicz 2021].

In spite of the common conviction that bees die immediately after stinging, it turns out that they can continue to live for another 18 to 114 hours. One insect injects on average 140 – 150 µg of venom, while the LD50 of bee venom is 2.8–3.5 mg of venom per kg of human body weight. Therefore, a non-allergic person with a weight of 60–70 kg has a 50% chance to die after being stung by 1000 – 1500 bees. Despite that, we need to be cautious because cases of death caused by 200 – 500 stings have been reported. The severity of the response also depends on age, body weight and the individual characteristics of the body [Pucca *et al.*, 2019]. The response of the body to venom is characterised by an allergic reaction manifesting itself in skin changes and problems with the respiratory, circulatory and digestive systems, whereas in the case of severe anaphylactic shock, in cerebral or myocardial ischemia. These effects are due to the presence of protein allergens in the venom composition, most of which exhibit enzymatic activity [David and Golden 2007; Wehbe *et al.*, 2019].

Apart from research on the negative effects of bee stings, science is also focused on the potential use of bee venom in treating chronic diseases.

In dermatology, bee venom finds application to alleviate symptoms of atopic dermatitis, from which 1–2% of adults and 20% of children suffer. A study conducted by You *et al.* [2016] involving application of an emollient with apitoxin gave a positive effect manifested in reduced eczema compared to the study group that used the same emollient without venom. Bee venom is also used in the cosmetics industry because it has been observed to have an effect on reducing skin discolourations and wrinkles. Purified, centrifuged and lyophilised venom in powder form is added to locally applied formulations [Łukasiewicz 2021]. This study revealed that melittin originating from *Apis mellifera* and *Apis florea* shows strong antitumor activity against breast and stomach cancer as well as melanoma. In spite of encouraging study results, there is apprehension concerning the activity of this protein against normal cells [Sangboonruang *et al.*, 2020; Yu *et al.*, 2020]. A common old age disease is arthrosis and accompanying inflammation. Adolapin isolated from bee venom has strong analgesic and anti-inflammatory properties that are probably associated with the inhibition of prostaglandin synthesis. The activity of phospholipase A2 taken from synovial fluid of a person suffering from rheumatoid arthritis is inhibited by combining it with melittin, which indicates reduced activity of the key inflammatory enzymes and this in turn means that bee venom exhibits potential therapeutic effects in this respect [Matysiak *et al.*, 2008].

6.3 STUDY RESULTS REGARDING HONEY QUALITY

Across the world, an increasingly greater attention is attached to honey quality. Organically produced honeys are gaining in importance, though as shown by our study [Table 6.2], both conventional lime honey, available in the market, and honey originating from an organic farm met all the required requirements. Regardless of the above

Table 6.2 The results of the analysis of selected parameters of conventional and organic linden honey (average value ± standard deviation) [Kwiatkowski *et al.*, 2022, unpublished data].

Parameter[*]	Conventional honey[**]	Organic honey[**]	Limit value[***]	HSD (p = 0.05)
I. Moisture [%]	17.7 ± 0.6 a	13.9 ± 0.4 b	< 20	3.1
II. Glucose and Fructose Content (sum of both) [g/100 g of honey]	62.3 ± 0.9 b	68.4 ± 0.7 a	> 50–60	5.2
III. Sucrose Content (g/100 g of honey)	3.2 ± 0.3 a	1.9 ± 0.2 b	< 5	0.7
IV. Diastase Number [Schade unit/g of honey]	23 ± 1 b	30 ± 2 a	> 8	4.0
V. Free Acids [mval/kg]	26.3 ± 0.8 a	18.5 ± 0.7 b	< 50	5.9
VI. Water Insoluble Solids Content/Ash [g/100 g of honey]	0.07 ± 0.01 a	0.06 ± 0.01 a	< 0.1	i.d.
VII. Content of 5-Hydroxy Methyl Furfural (HMF) [mg/kg of honey]	15.2 ± 0.9 a	5.6 ± 0.4 b	< 40	4.8

[*]the methods of determination of particular parameters, complied with Regulation of the Minister of Agriculture and Rural Development of January 14, 2009 establishing methods of analysis relating to evaluation of honey [Journal of Laws from 2009 No. 17 item 94 as amended] are as follows:

I. – refractometrically, in liquid honey. Determination procedure: 1) about 5 g of well-mixed honey was placed in a test tube and brought to a liquid state by heating in a water bath at a temperature of 40°C, 2) a few drops of honey were placed on the refractometer prism and covered with a dry matt plate, 3) the measurement was made and the index of refractive index was read to the fourth decimal place. The water content in honey (expressed in weight percent) corresponding to the determined refractive index was read from the table.

II. and III. – chromatographically, using an external standard. Determination procedure: 1) standards were prepared: about 2 g of glucose, 1.5 g of fructose and 0.25 g of sucrose were weighed out, placed in a beaker, and dissolved in about 40 ml of water, 2) 25 ml of methanol was pipetted into a 100 ml volumetric flask; 3) the standard solution was introduced into the flask with methanol and fill up with water to the mark; 4) honey samples were prepared: 5 g of honey was weighed and poured with 40 ml of water, 5) 25 ml of methanol was pipetted into a 100 ml volumetric flask and next the prepared honey solution was transferred quantitatively to the flask, the flask was filled up to the mark with water, 6) the standard solution and the honey solution were analyzed chromatographically at the following operating parameters of the chromatograph: a) carrier gas flow: 1.3 ml/min, b) column and detector temperature: 30°C, c) sample volume fed to the column: 10 μl.

IV. A unit of diastase activity was defined as the amount of enzyme that converts 0.01 g of starch to the specified endpoint in one hour, under temperature of 40°C. The results were expressed in Schade units per gram of honey. The diastatic activity of honey was determined by the photometric method, in which insoluble starch conjugated with a blue dye was used as a substrate. Due to hydrolysis of the starch by amylase, the water-soluble fragments of the starch chain bonded with the blue dye are formed. The absorbance of these fragments is measured photometrically at a wavelength of 620 nm.

V. based on pH measurement in an aqueous solution of honey, which was titrated to pH 8.3 with a 0.1 M sodium hydroxide solution. Determination procedure: 1) pH-meter was calibrated using pH 4.0, 7.0 and 9.0 buffers 2) 10 g of honey was placed in a 250 ml beaker and dissolved in 75 ml of carbon dioxide-free distilled water; 3) the samples was titrated with NaOH solution to a specified pH value, which was measured in a solution continuously mixed with use of magnetic stirrer.

VI. based on the amount of residue obtained during seepage of dissolved honey through the filter. Procedure of the determination: 1) about 25 g of honey was placed in a 150 ml beaker and diluted in about 100 ml of water at a temperature of about 70°C; 2) the honey solution was filtered through a soft filter paper, 3) the sample on the filter was washed with water at 70°C until the honey was completely rinsed out of the filter, 4) the filter with the residue was placed in a weighing container and dried for an hour in a laboratory dryer at 100°C; 4) after cooling in the desiccator, the filter with the residue was weighed with an accuracy of 0.0001 g; 5) The weighing container was transferred to a dryer at a temperature of 100°C for 30 minutes, cooled, and weighed again.

VII. by reversed-phase HPLC and UV detection, using standard solutions of 5-(hydroxymethyl) furan-2-carbaldehyde (HMF). HMF content was determined in a clear, filtered, aqueous solution of honey. The obtained signal was compared with those obtained for HMF standards of specified concentrations.

[**]results given in rows, marked with different letters (a-b) differ significantly (p = 0.05);

[***] according to the data given in Regulation ... , 2015, item 850.

i.d. – insignificant difference

statement, the study results prove that all the analysed quality parameters of lime honey (except for ash content) were significantly more favourable in the case of the organic honey than for the conventional one (which is confirmed by the calculated values of Tukey's Honestly Significance Difference (HSD), at p = 0.05).

Considering the results in detail, it should be stated that organic honey contained significantly less water (by 3.8 percentage points) compared to conventional one. However, both types of the examined honey met the requirements in terms of moisture content. The moisture content of the honeys tested by Chomaniuk *et al.* [2016] ranged from 16.8 to 27%, thus not all honeys met the limit values given in the relevant Regulation (2015). Also, the content of glucose and fructose in organic linden honey in the conducted study (68.4 g/100 g of honey) was significantly higher, by about 9%, compared to conventional honey. On the other hand, the content of sucrose in organic honey (1.9 g/100 g of honey) was significantly lower, by as much as 41%, than in that found in conventional one. Both parameters related to the content of sugars clearly confirm the higher quality of organic honey when compared to conventional one. On the basis of the results of Borawska *et al.* [2015], a higher sum of glucose and fructose in the linden honeys (nine samples from different sources) of 72.0 ± 0.2 g/100 g, but a lower content of sucrose – 0.6 ± 0.1 g/100 g was observed.

The high quality of organic linden honey was also confirmed by a significantly higher diastase number (by 24%), compared to conventional honey. The value of this parameter in the case of organic honey was 30 Schade units per g of honey. Diastase number determined for conventional linden honey by Godlewska & Świsłocka [2015] was about 17. In the presented research, the average value for conventional product was higher, and amounted to 23. An even greater difference in favor of organic linden honey (by about 30%) was noted in the case of the next analyzed parameter: acidity, i.e. the content of free acids. The average value of this parameter in organic honey was 18.5 mval/kg, while in conventional one it was 26.3 mval/kg. The lower values of this parameter indicate better honey quality. The results obtained in the conducted research were in the range of this parameter values given by Śliwińska & Bazylak [2011], which were from 8.5 to 26.5 mval/kg.

The water insoluble solids were determined as the ash content. The value of this parameter was at a comparable (statistically insignificant difference) level in both examined types of honey. The content of this ingredient did not exceed the legal limit of 0.1 g/100 g of honey in any the examined samples.

The content of 5-hydroxymethylfurfural (HMF) in organic linden honey was low (5.6 mg/kg of honey), which is another feature confirming its high quality. The content of HMF in the conventional linden honey was ca. 2.7 times higher. HMF is formed in honey due to dehydration of simple sugars and concentration of this compound depends on storage time and temperature. High HMF values, especially exceeding 40 mg/kg, indicate the adulteration of honey, e.g. by the addition of sugar syrups. The conventional linden honey studied by Godlewska & Świsłocka [2015] contained slightly over 2 mg of HMF/kg of the product, which indicates its high quality, similar to the organic honey assessed in the presented study.

Considering the data given in Table 6.3, it was noted that values of all the examined parameters of acacia honey, both the conventional and organic ones, did not exceed the limits given in Regulation ... 2015, item 850, which confirms the relatively high quality of acacia honey.

However, the results of the study showed that in the case of acacia honey, fewer statistically significant differences between the average values of particular parameters determined in conventional and organic products were observed. Diastase number, acidity of honey and the content of ash were similar in both types of honey. It follows that the tested organic acacia honey did not stand out in terms of quality as clearly as organic linden honey from conventional one. Nevertheless, parameters such as: water content, simple sugars and sucrose content, and HMF content were significantly in favor of organic acacia honey over the conventional one. This is confirmed by the NIR values (p = 0.05) quoted in Table 6.3.

Table 6.3 The results of the analysis of selected parameters of conventional and organic acacia honey [Kwiatkowski *et al.*, 2022, unpublished data].

Parameter[*]	Conventional honey[**]	Organic honey[**]	Limit value[***]	HSD (p = 0.05)
I. Moisture [%]	18.5 ± 0.6 a	14.8 ± 0.4 b	< 20	3.0
II. Glucose and Fructose Content (sum of both) [g/100 g of honey]	65.6 ± 0.7 b	70.4 ± 0.5 a	> 50–60	4.1
III. Sucrose Content (g/100 g of honey)	4.5 ± 0.4 a	2.5 ± 0.3 b	< 5	0.9
IV. Diastase Number [Schade unit/g of honey]	13 ± 1 b	15 ± 2 a	> 8	n.s.
V. Free Acids [mval/kg]	16.4 ± 0.5 a	14.8 ± 0.6 b	< 50	n.s.
VI. Water Insoluble Solids Content/Ash [g/100 g of honey]	0.09 ± 0.02 a	0.08 ± 0.01 a	< 0.1	i.d.
VII. Content of 5-Hydroxy Methyl Furfural (HMF) [mg/kg of honey]	21.0 ± 1.1 a	10.2 ± 0.6 b	< 40	5.1

Footnotes as below Table 6.2.

The water content in conventional honey was at a similar level as in the study by Jasińska *et al.* [2020], in which the values of this parameter in commercially available acacia honey (five different samples) were in the range of 17–19%. The content of sucrose in the honey examined by these authors was in the range of 1.5–5 g/100 g, hence, the results obtained in presented study were in this range; while the total acidity in the honey examined by Jasińska *et al.* [2020], which ranged from 2 to 7 mval/kg, was much lower than that measured by the authors of this paper. The ash content in the products tested in the conducted study was at the same level as that determined by Majewska *et al.* [2015] in acacia honeys which were purchased directly from the beekeeper – which was 0.08%, and the values of the diastase number were similar to the value given by Godlewska & Świsłocka [2015]. In the case of HMF, the content of this compound in the commercially available acacia honey examined by Śliwińska *et al.* [2012] was high (29.4–31.7 mg/kg). This value indicates a worser quality of these honeys even compared to the conventional honey examined in the presented study (Table 6.3).

Comparing the quality parameters of both analyzed honeys (Tables 6.1 and 6.2), it should be noted that regardless of the honey production method (conventional or organic), linden honey contained on average less water (13.9–17.7%), less sucrose (1.9–3.2 g/100 g honey), less HMF (5.6–15.2 mg/ kg of honey) when compared to acacia honey, but the content of simple sugars (glucose and fructose) was similar in these honeys. However, linden honey was characterized by a higher amount of free acids (18.5–26.3 mval/kg) than acacia honey (14.8–16.4 mval/kg).

Summing up, the conducted study showed that organic honey production generally results in better honey quality than conventional one. Among the two compared products, linden honey showed a more favorable response to the organic production system, which was reflected in the values of most the parameters analyzed.

Not only bees, but also beekeepers have a significant impact on the value of honey. They are responsible for the quality of the product, starting from honey harvesting to the sale. The most common defects occurring during the production of honey are: fermentation, overheating, and stratification. It is recommended to locate the apiary away from roads and chemical plants. Melliferous plants are the best neighborhood for an apiary. Preserving the health properties of honey also depends on the consumer. The most common mistakes made after purchase are improper storage and heating to a temperature above 40°C [Szwedziak et al., 2017]. Besides the botanical and geographical origin, the physical and chemical properties of honey depend on transport, packaging and storage conditions, as well as external factors, e.g. climatic and weather conditions [Żak et al., 2017].

Consonni et al. [2019] note that in recent decades, organic farming has had a major impact on social and political thinking, including honey production. There are not many studies focusing on the comparison of organic and conventional honeys but the few that have been carried out show the advantage of organic over conventional honey, for example due to more favorable antioxidant properties.

Obey et al. [2022] claimed that the commercially available organic honeys derived from different countries exhibit antimicrobial activity against numerous human bacterial pathogens. It has been shown that organic honeys can be used as a means of preventing infections. It is also possible to use them to create new antibacterial food additives.

According to Polak-Śliwińska & Tańska [2021] the organic honeys were characterized by darker color and intensive yellowness, higher content of flavonoids and vitamin C, but lower concentration of HMF when compared to the conventional ones. On the other hand, the conventional honeys contained slightly higher phenolics, and their antioxidant activity was also higher, regardless of the extraction solvent employed during the test.

Due to the abundance of antioxidants in honey, recent research suggests that it provides protection against a variety of conditions, including diabetes mellitus, respiratory, gastrointestinal, cardiovascular, nervous system diseases as well as cancer treatment. It was discovered that the organoleptic properties of honey may constitute an initial indication of its higher antioxidant potential. According to Polak-Śliwińska and Tańska [2021], in general, honeys with a light color, a delicate aroma, and a mild flavor were found to have a lower total phenolic content and lower antioxidant activity. In contrast, brown and spice honeys were characterized by high concentration of the aforementioned compounds.

The organoleptic characteristics of Serbian monofloral (acacia) and multifloral (meadow) honeys also differ significantly, as reported by Popov-Ralji et al. [2015]. They observed that differences in properties of honey stems from varying botanical and geographical origins, chemical compositions, weather, beekeeping practices, etc. Silvano et al. [2014] studied the sensory and physicochemical properties of honey from various parts of the Buenos Aires province. They believe that physicochemical parameters could be used to classify honeys according to their geographic origin, whereas sensory properties were not reliable as predictors.

Perez-Arquillue *et al.* [1994] found a sucrose content below 1 g/100 g in a variety of Spanish honey samples of varying botanical origin. They observed that six samples did not contain sucrose could be attributed to the fact that these honeys were obtained from nectars that only contained fructose and glucose, since sucrose was completely converted into monosaccharides before the nectars were secreted.

6.4 SURVEY RESEARCH ON THE CONSUMPTION OF HONEY AND BEE PRODUCTS IN POLAND

The study was conducted using the diagnostic survey method (internet survey technique with a proprietary survey form) among randomly selected people (anonymous potential consumers of honey and other bee products in Poland). The research sheet contained 17 questions, both closed and open-ended. Closed questions were single or multiple choice. The study, which was conducted in 2022, covered 700 people of different ages. Women dominated among the respondents (Figure 6.1a). The age analysis of the respondents showed that the largest number of respondents were over 45 (52.9%) and 36–45 (27.1%). Other age groups accounted for a total of 18% (Figure 6.1b).

Most of the respondents (over 60%) were city dwellers (Figure 6.1c). More than half (53%) of the respondents had higher education. The rest were dominated by people with

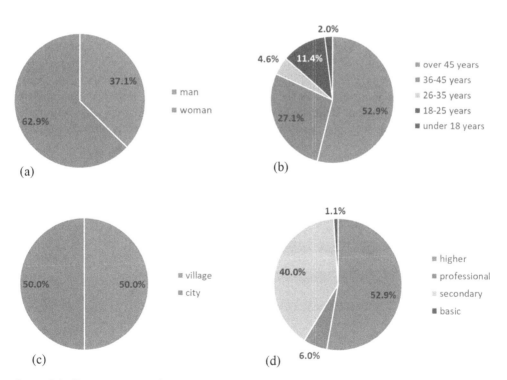

Figure 6.1 Characteristics of respondents by: (a) gender, (b) age, (c) place of residence, (d) education of respondents (% of respondents).

secondary education. Only about 7% of the respondents had primary or vocational education (Figure 6.1d).

The survey results showed that 35.7% of the respondents consumed honey several times a year, and 31.4% several times a month. About 21% of respondents indicated that they consume honey more often (once a week or every day), while the rest (11.4%) used honey only once a month (Figure 6.2a). The research results showed that 50% of the respondents consume less than 1 kg of honey per year, and nearly 43% consume from 1 to 4 kg, the rest used annually a more honey (Figure 6.2b). Most of the respondents correspond to the eating habits adopted in Poland, because according to the data of the Central Statistical Office, there is only 0.36 kg of honey per inhabitant of Poland per year. This is about half the average amount of honey consumed per capita in Europe [Borowska 2016].

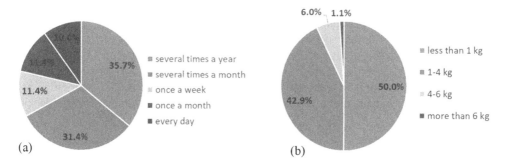

Figure 6.2 Characteristics of the respondents in terms of: (a) frequency (b) amount of honey consumed per year (% of respondents).

The taste is the main reason of which the respondents consume honey. This factor was indicated by as many as 63% of the respondents. In addition, honey use as a dietary supplement and a product with medicinal or nutritional values were also important (Figure 6.3). Habit was the least significant factor influencing the decision to consume honey.

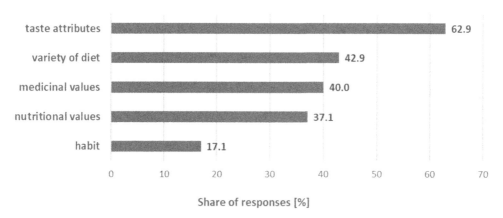

Figure 6.3 Factors determining the consumption of honey by respondents (% of respondents).

Most respondents indicated that they take into account the type (nearly 65%) and consistency of honey (50%) making decision on purchasing the product. According to the respondents, the taste, color and smell of honey were less important (Figure 6.4). Low share of indication the taste and smell may result from the fact that it is rarely possible to taste honey when buying it.

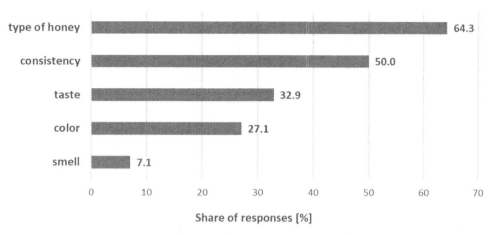

Figure 6.4 Honey features influencing its purchase by respondents (% of respondents).

The results of the survey showed that while purchasing honey, most consumers do not pay attention to the organic nature of its production. For more than half of the respondents, the origin of the honey did not matter. Over 1/3 of the respondents bought honey from conventional beekeeping farms, and only about 16% chose honey from organic beekeeping farms (Figure 6.5).

Taking into account the type of honey, multiflower honey was the most preferred by the respondents (about 46% of respondents chose it). The following honeys were

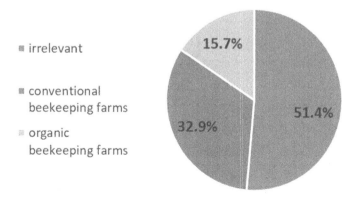

Figure 6.5 Importance of honey origin for consumers (% of respondents).

significantly less popular: linden, honeydew, rapeseed and buckwheat (Figure 6.6). Honeys produced from phacelia, acacia and heather were selected the least frequently.

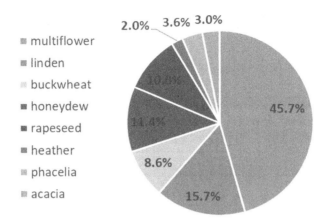

Figure 6.6 Preferences of respondents regarding the types of honey (% of respondents).

The main factors which influence the purchase of honey by respondents were: high quality of honey (about 26%) and price (almost 23%). The respondents also paid attention to the producer, habits of using a particular type of honey, and honey availability (Figure 6.7). On the other hand, the attractiveness of the packaging, the manufacturer's quality certificate and advertising had no significant impact on the purchase decision. These factors were indicated in total by only 5% of the respondents.

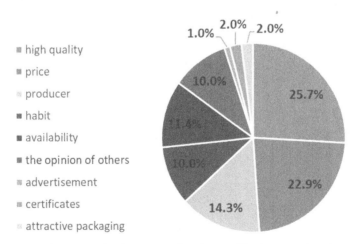

Figure 6.7 Factors influencing the decision to buy honey (% of respondents).

The research showed that honey and bee products are most often purchased directly from the beekeeper. This source of purchase was indicated by as many as half of the respondents. The market/bazaar as source of honey came second, and the supermarket the third one. Buying honey over the Internet was seldom practiced (Figure 6.8).

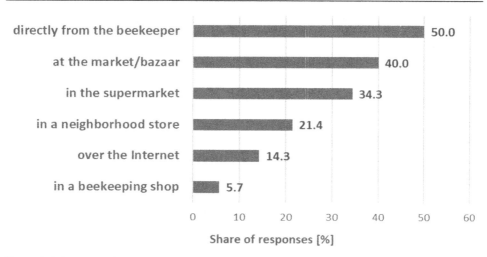

Figure 6.8 Sources of purchase of honey and bee products (% of respondents).

The respondents most often use honey to sweeten drinks (about 75% of responses), prepare meals (66%) and for prophylactic purposes (almost 49%). The respondents also indicated the use of honey for the production of alcoholic beverages and for cosmetic purposes (Figure 6.9).

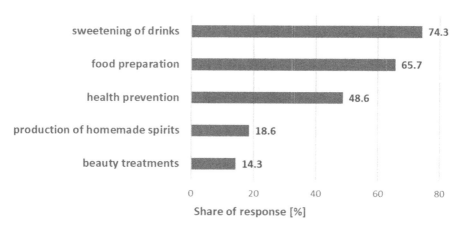

Figure 6.9 Purpose of honey in the household (% of respondents).

Respondents were familiar with bee products, and the most popular in their opinion are honey, propolis, pollen and wax. Respondents use bee products such as bee bread and royal jelly less frequently. Due to the specificity of use, bee venom is the least popular (Table 6.4).

According to the data presented in Figure 6.10, the respondents often used bee products. Only less than 6% of respondents did not use them at all. The rest, on the other hand, used bee products primarily for prophylactic purposes (over 60% of indications), followed by health and cosmetic purposes (Figure 6.10).

Table 6.4 The most popular bee products in the opinion of respondents.

Bee products	Share of response (%)
Honey	70.3
Propolis (bee glue)	49.7
Flower pollen	40.0
Beeswax	35.7
Bee bread	30.2
Royal jelly	21.3
Bee venom	14.2

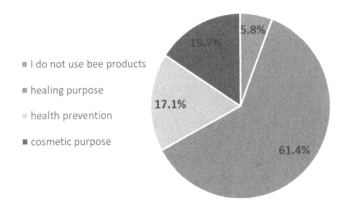

Figure 6.10 Purposes of using bee products (% of respondents).

To sum up, the importance of organic methods of bee breeding should be emphasized. These methods contribute to an increased or at least comparable nutritional and health-promoting value of honey compared to those produced in conventional beekeeping farms.

The analyzed honeys, regardless of whether the production was conventional or organic, were of good quality and met the legal requirements specified in the relevant standards. However, organic production resulted in improved honey properties, such as higher glucose and fructose content, higher diastase number, lower moisture, as well as sucrose, free acids, and HMF contents.

Honey and other bee products are commonly used as a dietary supplement. The conducted surveys showed that the respondents often consume honey, mainly because of its taste, but they do not pay attention to the type of the beekeeping farming.

Conclusions

The safe future of man on the Earth depends to a large extent on the availability of healthy food and the quality of the environment. The growth of the human population forces an increase in food production, including that of animal origin, which results in the development of cropping systems and breeding farms on an industrial scale. Such production, in turn, is associated with many threats to the environment, including unfavorable climatic or hydrological changes. That is why it is so important to change agricultural systems, aimed at sustainable agricultural production, so as to ensure food security, protection of soil, water and climate on a global scale. Organic farming systems are moving in this direction. However, such farms constitute a small percentage of the share in the total pool of agricultural holdings. Conventional agriculture still dominates, within which, however, farming in a sustainable and environmentally safe manner, which is a "mitigated form" of conventional agriculture, is becoming more and more common.

The current state of knowledge does not allow clearly stating how farming system affects the quality of agricultural products and food produced in them, and whether food from organic farms is more valuable than that derived from conventional farms.

This review of the literature aimed at comparing the quality of organic and conventional agri-food products indicates that the tested products from both farming systems met the quality requirements set out in the applicable regulations in terms of the parameters covered by the research; at the same time, the results of the research indicate that organic agri-food products are characterized by better nutritional and health-promoting properties compared to those obtained from conventional farms.

Research on the impact of the type of farming system on food quality conducted by researchers from various countries, as well as those presented in this monograph, confirm the high nutritional and health-promoting values of plant-derived products, such as millet groats, canned green peas, edible carrot roots, zucchini fruits, apples and bananas, as well as products of animal origin – two types of honey – obtained organically. The comparative analysis showed a higher content of total solid, carbohydrate and protein, better antioxidant properties, as well as higher content of vitamins and microelements, which indicates a potentially better effect on human health and wellness. Simultaneously, the research shows that the conventional food tested met the applicable quality standards.

The survey conducted among honey consumers showed, however, that despite the fact that a large group of respondents consume this product, mainly because of its taste, few people pay attention to its source of origin. Nevertheless, as indicated by the results

DOI: 10.1201/9781003380771-7

of research on the quality parameters of honeys conducted by the authors of this monograph and other researchers, the honeys from organic beekeeping farms show more favorable features compared to those from conventional ones. Significant differences were found for a number of nutritional parameters of linden and acacia honey, such as: higher diastase number, simple sugar content, lower acidity, content of moisture, sucrose and water-insoluble impurities.

When buying food products, it is worth paying attention to their origin, because their nutritional and/or health-promoting values in products obtained from farms run according to the principles of organic farming may be higher.

It can be predicted that organic plant production, as well as organic beekeeping, due to the high quality of the products obtained, will enjoy increasing popularity among conscious consumers. Organic farming produces better quality food. However, there are doubts whether this is a method that can become popular enough to replace conventional agriculture, which after all appeared to respond to the needs of the growing human population. The main problem is quantitative limitations related to crop yield. However, this issue is a field of wide discussion, where opposing views clash, and is a topic for a separate study.

References

AACC. *Approved Method of the AACC*, 9th ed.; American association of cereal chemists: Saint Paul, MN, USA, 1995.

Abdussalam-Mohammeda A., Alia A.Q., Errayes A.O. Green chemistry: Principles, applications, and disadvantages. *Chem. Methodol.* 2020, 4, 408–423. doi: 10.33945/SAMI/CHEMM.2020.4.4

Achremowicz B., Kuczyński A., Puchalski C., Wiśniewski R., Kaszuba J. Comparison of quality of oat flakes from organic and conventional production. *Science Nature Technologies – Nauka Przyroda Technologie.* 2016, 10, 3,28. doi: 10.17306/J.NPT.2016.3.28.

Adamczyk M., Rembiałkowska E. The comparative assessment of some selected quality characteristics of apples from the organic and conventional production. *Żywność. Nauka. Technologia. Jakość/Food. Science. Technology. Quality.* 2005, 12, 2, (43) Supl., 9–21.

Ahmad S., Campos M.G., Fratini F., Altaye S.Z., Li J. New insights into the biological and pharmaceutical properties of royal jelly. *Int. J. Mol. Sci.* 2020, 8; 21(2), 382. doi: 10.3390/ijms21020382.

Aktar M.W., Sengupta D., Chowdhury A. Impact of pesticides use in agriculture: their benefits and hazards. *Interdiscip. Toxicol.* 2009, 2(1), 1–12. doi: 10.2478/v10102-009-0001-7. PMID: 21217838; PMCID: PMC2984095.

Alemayehu T., Assogba G.M., Gabbert S., Giller K.E., Hammond J., Arouna A., Dossou-Yovo E.R., van de Ven G.W.J. Farming systems, food security and farmers' awareness of ecosystem services in Inland Valleys: A study from côte d'Ivoire and ghana. *Front. Sustain. Food Syst.* 2022 Sec. Land, Livelihoods and Food Security

Ali A.M., Kunugi H. Apitherapy for age-related skeletal muscle dysfunction (sarcopenia): A review on the effects of royal jelly, propolis, and bee pollen. *Foods* 2020, 9, 1362. https://doi.org/10.3390/foods9101362

Ali B. M. M., Ghoname N. F., Hodeib A. A., Elbadawy M. A. Significance of topical propolis in the treatment of facial acne vulgaris. *Egyptian Journal of Dermatology and Venerology* 2015, 1, 35, 29–36.

Alix A., Capri E. Chapter one – modern agriculture in europe and the role of pesticides, Editor(s): E. Capri, A. Alix, *Advances in Chemical Pollution*, Environ. Management Prot. *Elsevier. 2018, 2, 1–22. https://doi.org/10.1016/bs.apmp.2018.04.001.*

Allen T. Ecosystem sustainability, agricultural biodiversity and diet quality: A system and cultural updates. *Public Health Nutrition* 2013, 14(12A).

America's animal factories: How states fail to prevent pollution from livestock waste. 1988. Washington: Natural Resources Defence Council, Clean Water Council.

Amvrazi E.G. Fate of pesticide residues on raw agricultural crops after postharvest storage and food processing to edible portions. In *Pesticides-Formulations, Effects, Fate*; IntechOpen: London, UK, 2011.

Anastas P., Eghbali N. Green chemistry: Principles and practice. *Chem. Soc. Rev.* 2010, 39, 301–312. doi: 10.1039/B918763B

Anastassiades M., Lehotay S., Štajnbaher D. Quick, easy, cheap, effective, rugged, and safe (QuEChERS) approach for the determination of pesticide residues. In *Pesticide Protocols*; Humana Press: Totowa, NJ, USA, 2002.

Anjum S.I., Ullah A., Khan K.A., Attaullah M., Khan H., Ali H., Bashir M.A., Tahir M., Ansari M.J., Ghramh H.A., Adgaba N., Dash C.K. Composition and functional properties of propolis (bee glue): A review. *Saudi J. Biol. Sci.* 2019, 26(7), 1695–1703. doi: 10.1016/j.sjbs.2018.08.013.

Avilova S.V., Kornienko V.N., Gryzunov A.A., Vankova A.A. An effect of storage and transportation temperature on quantitative composition of microflora of plant products. *Food Syst.* 2019, 2, 4, 42–47.

Bach-Faig A., Berry E., Lairon D. *et al.* Mediterranean diet pyramid today. *Science and cultural updates.* Public Health Nutrition 2011, 14(12A).

Bajwa U., Sandhu K.S. Effect of handling and processing on pesticide residues in food- a review. *J. Food Sci. Technol.* 2014, 51(2), 201–20. doi: 10.1007/s13197-011-0499-5.

Bakour M., Laaroussi H., Ferreira-Santos P., Genisheva Z., Ousaaid D., Teixeira, J.A., Lyoussi B. Exploring the palynological, chemical, and bioactive properties of non-studied bee pollen and honey from morocco. *Molecules* 2022, 27, 5777. https://doi.org/10.3390/molecules27185777

Barzman M., Barberi P., Birch A.N.E., Boonekamp P., Dachbrodt-Saaydeh S., Graf B. *et al.* Eight principles of integrated pest management. *Agron. Sustain. Dev.* 2015, 35, 1199–1215. doi: 10.1007/s13593-015-0327-9

BCFN Double Pyramid 2012. Enabling sustainable food choices. *Parma*, 2012a.

BCFN Double Pyramid 2012. Food styles and environmental impact. *Parma*, 2012b.

Benbrook C., Xin Z., Jaime Y., Neal D., Preston A.: New evidence confirms the nutritional superiority of plant-based organic foods. *State of Science Review: Nutritional Superiority of Organic Food* 2008.

Benkebila, N. 2015. *Agroecology, Ecosystems and Sustainability*. Boca Raton, London, New York, Leiden: CRC Press, Taylor & Francis Group, A Balkema Book.

Bhadekar R., Pote S., V. Tale V., Nirichan B. Developments in analytical methods for detection of pesticides in environmental samples. *American Journal of Analytical Chemistry* 2011, 2, 8A, 1–15. doi: 10.4236/ajac.2011.228118.

Biancolillo A., Boqué R., Cocchi M., Marini F. *Data Fusion Strategies in Food Analysis*. Elsevier 2019, 31, 271–310.

Biehler E., Mayer F., Hoffmann L., Krause E., Bohn T. Comparison of 3 spectrophotometric methods for carotenoid determination in frequently consumed fruits and vegetables. *J. Food Sci.* 2010, 75(1), C55–61. doi: 10.1111/j.1750-3841.2009.01417.x.

Bobrowska-Korczak B., Wójcik A., Tokarz A. The content of vitamin c in vegetables and fruits from ecological and conventional production. *Bromatol. Chem.* Toksykol. 2016, 49, 3, 225–228.

Bocho-Janiszewska A., Sikora A., Rajewski J., Łobodzin P. Application of royal jelly in moisturizing creams. *Pol J Cosmet.* 2013, 16, 4, 314–320.

Borawska M., Arciuch L., Puścion-Jakubik A., Lewoc D.: Zawartość cukrów (fruktozy, glukozy, sacharozy) i proliny w różnych odmianach naturalnych miodów pszczelich [Content of sugars (fructose, glucose, sucrose) and proline in different varieties of natural bee honey]. *Probl. Hig. Epidemiol.* 2015, 96, 4, 816–820. [in Polish]

Borch A., Kjærnes U. Food security and food insecurity in Europe: An analysis of the academic discourse (1975–2013). *Appetite* 2016, 103, 137–147. http://dx.doi.org/10.1016/j.appet.2016.04.005

Borowska A., Production, consumption and foreign trade of honey in Poland in the years 2004 to 2015, *Roczniki Naukowe Ekonomii Rolnictwa I Rozwoju Obszarów Wiejskich*, T. 103, z. 4, 2016.

Botitsi H., Tsipi D., Economou A. Current legislation on pesticides. In *Applications in high resolution mass spectrometry* (pp. 83–130). Elsevier 2017. doi:10.1016/B978-0-12-809464-8.00004-X

Bułkowska, K., Gusiatyn, Z.M., Klimiuk, E., Pawłowski, A., Pokój, T. 2016. *Biomass for biofuels. Boca Raton*, London, New York, Leiden: CRC Press, Taylor & Francis Group, A Balkema Book.

Burchat C.S., Ripley B.D., Leishman P.D., Ritcey G.M., Kakuda Y., Stephenson G.R. The distribution of nine pesticides between the juice and pulp of carrots and tomatoes after home processing. *Food Addit Contam.* 1998;15:61–71.

Burchi, F., de Muro, P. From food availability to nutritional capabilities: Advancing food security analysis. *Food Policy* 2016, 60, 10–19. http://dx.doi.org/10.1016/j.foodpol.2015.03.008.

Burg V., Rolli C., Schnorf V., Scharfy D., Anspach V., Bowman G. Agricultural biogas plants as a hub to foster circular economy and bioenergy: An assessment using substance and energy flow analysis, *Resources, Conservation and Recycling*, 2023, 190, 106770, https://doi.org/10.1016/j.resconrec.2022. 106770.

Burney, J.A., Davis, S.J., Lobell, D.B. 2010. Greenhouse gas mitigation by agricultural intensification. *Proc. of the National Academy of Sciences of the United States of America* 107(26): 12052–12057.

Butnariu M., Sarac I. Functional food. *Int. J. Nutr.* 2019, 3(3), 7–16. https://doi.org/10.14302/issn.2379-7835.ijn-19-2615

Cabras P, Angioni A. Pesticide residues in grapes, wine and their processing products. *J Agric. Food Chem.* 2000, 48, 967–973.

Campbell, M.N. 2008. Biodiesel: algae as a renewable source for liquid fuel. *Guelph Engineering Journal* 1: 2–7.

Can Z., Yildiz O., Sahin H., Turumtay E.A., Silici S., Kolayli S. An investigation of Turkish honeys: Their physico-chemical properties, antioxidant capacities and phenolic profiles, *Food Chem.* 2015, 180, 133–141. https://doi.org/10.1016/j.foodchem.2015.02.024.

Capoci I.R.G., Bonfim-Mendonça Pde S., Arita G.S., de Araújo Pereira R.R., Edilaine Lopes Consolaro M., Bruschi M.L., Negri M., Estivalet Svidzinski T.I. Propolis is an efficient fungicide and inhibitor of biofilm production by vaginal *Candida albicans. Evidence-Based Complementary and Alternative Medicine*, Article ID 287693, 9 pp., 2015. https://doi.org/10.1155/2015/2876932015.

Cardoso P.C., Tomazini A.P.B., Stringheta P.C., Ribeiro S.M.R., Pinheiro-Sant'Ana H.M. Vitamin C and carotenoids in organic and conventional fruits grown in Brazil. *Food Chem.* 2011, 126, 2, 411–416. https://doi.org/10.1016/j.foodchem.2010.10.109.

Çatak J. Determination of niacin profiles in some animal and plant based foods by high performance liquid chromatography: Association with healthy nutrition. *J. Anim. Sci. Technol.* 2019, 61(3),138–146. https://doi.org/10.5187/jast.2019.61.3.138

Chang K.H., Wu R.Y., Chuang K.C., Hsieh T.F., Chung R.S. Effects of chemical and organic fertilizers on the growth, flower quality and nutrient uptake of *Anthurium andreanum*, cultivated for cut flower production. *Sci Hortic-Amsterdam.* 2010, 125, 434–441.

Chavarri M.J., Herrera A., Arino A. The decrease in pesticides in fruit and vegetables during commercial processing. *Int. J. Food Sci. Technol.* 2005, 40, 205–211.

Chazelas E., Deschasaux M., Srour B. *et al.* Food additives: Distribution and co-occurrence in 126,000 food products of the French market. *Sci. Rep.* 2020, 10, 3980. https://doi.org/10.1038/s41598-020-60948-w

Chhetri P., Gudade A., Bhattari N. K.: Need of organic plant breeding system in organic farming. *Popular Kheti.* 2013, 1(4), 231–234.

Chmielewska M., Tys J., Petkowicz J., Petkowicz B. Food – first, to do no harm (a review). *Acta Agrophys.* 2018, 25(1), 17–34. DOI: https://doi.org/10.31545/aagr0002

Chopra A. K., Sharma M. K., Chamoli S. Bioaccumulation of organochlorine pesticides in aquatic system-an overview. *Environ. Monit.* 2011, Assess. 173.

Chotyakul N., Pateiro-Moure M., Saraiva J. A., Torres J. A., Pérez-Lamela C. Simultaneous HPLC–DAD quantification of vitamins A and E content in raw, pasteurized, and UHT cow's milk and their changes during storage. *Eur. Food Res. Tech.* 2014, 238(4), 535–547. doi:10.1007/s00217-013-2130-7

Christaki E., Bonos E., Giannenas I., Florou-Paneri P. Aromatic plants as a source of bioactive compounds. *Agriculture* 2012, 2, 228–243. https://doi.org/10.3390/agriculture2030228

Chung S.W. How effective are common household preparations on removing pesticide residues from fruit and vegetables? A review. *J. Sci. Food Agric.* 2018, 98, 2857–2870.

Code of Federal Regulation, Title 40 – Protection of Environment, Chapter I – Environmental protection agency, Subchapter E – Pesticide programs, Part 156 – Labeling Requirements for Pesticides and Devices, Subpart A – General Provisions, § 156.10 Labeling requirements, https://www.ecfr.gov/current/title-40/chapter-I/subchapter-E/part-156?toc=1 [last accessed: 2023.02.15].

Commission Implementing Regulation (EU) 2021/1165 of 15 July 2021 Authorising Certain Products and Substances for Use in Organic Production and Establishing their Lists (OJ L 253/39. Available online: https://eur-lex.europa.eu/legal-content/EN/TXT/PDF/?uri=CELEX:32021R1165 [last accessed 12 March 2023].

Committee on Climate Change, 2018, Biomass in a Low-Carbon Economy. London: Thecccuk.

Consonni R., Bernareggi F., Cagliani L.R. NMR-based metabolomic approach to differentiate organic and conventional Italian honey, *Food Control* 2019, 98, 133–140. https://doi.org/10.1016/j.foodcont.2018.11.007

Cooney, C.M. 2006. Sustainable agriculture delivers the crops. *Environ. Sci. Technol.* 15: 1091–1092.

Cornara L., Biagi M., Xiao J., Burlando B. Therapeutic properties of bioactive compounds from different honeybee products. *Front. Pharmacol.* 2017, 28, 8, 412. doi: 10.3389/fphar.2017.00412.

Cwalina-Ambroziak B. Effects of different farming systems and crop protection strategies on the health status and yield of carrots *Daucus carota* L. ssp. *sativus. Acta Sci. Pol. Hortorum Cultus* 2022, 21, 2, 3–17. https://doi.org/10.24326/asphc.2022.2.1

Czeczot H., Majewska M. Cadmium – exposure and its effects on health. *Farmacja Polska* 2010, 66(4), 243–250.

Daioglou, V., Doelman, J.C., Wicke, B., Faaij, A., van Vuuren, P. 2019. Integrated assessment of biomass supply and demand in climate change mitigation scénarios, *Global Environ. Change* 54: 88–101.

Dangour A.D., Dodhia S.K., Hayter A., Allen E., Lock K., Uauy R. Nutritional quality of organic foods: a systematic review. *Am. J. Clinic. Nutrition* 2009, 90, 3, 680–685, https://doi.org/10.3945/ajcn.2009.28041

Darnhofer I., Bellon S., Dedieu B., Milestad R. Adaptiveness to enhance the sustainability of farming systems. A review. *Agronomy for Sustainable Development* 2010, 30 (3), 545–555. ff10.1051/agro/2009053ff. ffhal-00886511f

David B. K., Golden M. D.: Insect sting anaphylaxis. *Immunology and Allergy Clinics of North America,* 2007, 27, 2. doi: 10.1016/j.iac.2007.03.008

de Groot R. 2011. What are ecosystem services? *Treatise on Estuarine and Coastal Science* 12: 15–34.

de Ponti T., Rijk B., van Ittersum M. K. The crop yield gap between organic and conventional agriculture. *Agric. Syst.* 2012, 108, 1–9.

De Vires, B.J.M., 2013. *Sustainability Science.* New York: Cambridge University Press.

Deadman M.L. Sources of pesticide residues in food: Toxicity, exposure, and risk associated with use at the farm level. *In: Pesticide Residue in Foods.* Springer Int. Publ. 2017, 7–35. doi:10.1007/978-3-319-52683-6_2

Dickie, A., Streck, C., Roe, S., Zurek, M., Haupt, F. and Dolginow, A. 2014. Strategies for Mitigating Climate Change in Agriculture: *Recommendations for Philanthropy.* https://www.agriculturalmitigation.org [19.02.2023].

Ditlevsen K., Sandøe P., Lassen J.: Healthy food is nutritious, but organic food is healthy because it is pure: The negotiation of healthy food choices by danish consumers of organic food. *Food Qual.* Prefer. *2019, 71, 46–53.*

Dixon J.L., Stringer L.C., Challinor A.J. Farming system evolution and adaptive capacity: Insights for adaptation support. *Resources* 2014, 3, 182–214. https://doi.org/10.3390/resources3010182

Dranca F., Ursachi F., Oroian M. Bee Bread: Physicochemical Characterization and Phenolic Content Extraction Optimization. *Foods* 2020, 9, 1358. https://doi.org/10.3390/foods9101358

Drobnica L., Cebulak T., Pieczona W. Nourishment and chronic non-contagious diseases in the opinion of consumers of the non-conventional food. *Żywność. Nauka. Technologia. Jakość.* 2007, 6 (55).

Dumitru C.D., Neacsu I.A., Grumezescu A.M., Andronescu E. Bee-Derived products: Chemical composition and applications in skin tissue engineering. *Pharmaceutics* 2022, 14, 750. https://doi.org/10.3390/pharmaceutics14040750

EC 2020. *European Green Deal: Commission Prepares New Initiatives to Boost the Organic Farming Sector.* 2020a. Retrieved from https://ec.europa.eu/commission/ presscorner/detail/en/IP_20_1548 (20.11.2022).

EC 2020. *How the Future CAP will Contribute to the EU Green Deal.* 2020b. https://ec.europa.eu/info/sites/info/files/food-farming-fisheries/sustainability_a nd_natural_resources/documents/factsheet-how-cap-contributes-to-green-deal_ en.pdf

EC. 2001. *Biomass,* https://energy.ec.europa.eu/topics/renewable-energy/bioenergy/biomass_en [21.02.2023].

EC. 2019. *2020* Climate & Energy Package. *Brussels: European Commission. 2019.*

EC. 2022a. *Organic Farming,* https://agriculture.ec.europa.eu/farming/organic-farming_en [19.02.2023].

EC. 2022c. *Final Communication from the Commission to the European Parliament, The European Council, The Council of The European Economic and Social Committee and the Committee of the Regions, REPowerEU Plan.* EC: Brussels.

EC.a2022b. *Organic Farming Statistics,* https://ec.europa.eu/eurostat/statistics-explained/index. php?title=Organic_farming_statistics#Total_organic_area [19.02.2023].

Esnouf, C., Russel, M., Bricas, N. 2013. *Food System Sustainability.* New York: Cambridge University Press.

Esti M., Cinquanta L., Sinesio F., Moneta E., Di Matteo M. Physicochemical and sensory fruit characteristics of two sweet cherry cultivars after cool storage, *Food Chemistry* 2002, 76, 4, 399–405. https://doi.org/10.1016/S0308-8146(01)00231-X

European Commission. *Report from the Commission to the European Parliament and the Council Regarding the Use of Additional Forms of Expression and Presentation of the Nutrition Declaration*; European Commision: Brussels, Belgium, 2020, pp. 25.

European Economic and Social Committee. *NAT/596 Integrated production in the European Union,* Bruksela, 2014.

European Environment Agency. 2021. *EU Achieves 20-20-20 Climate Targets, 55 % Emissions Cut By 2030 Reachable with More Efforts and Policies.* EEA, Brussels.

European Union Official Website, Organic Farming, https://agriculture.ec.europa.eu/farming/organic-farming/organics-glance_en [last accessed 2 May 2023]

Eurostat 2020. Organic farming statistics. *In: Statistics Explained* 2020b. https://ec.europa.eu/eurostat/statisticsexplained/index.php?title=Organic_farming_statistics. Accessed 10 Jan 2023

Eurostat 2020. *Sales of Pesticides in the EU.* 2020a. https://ec.europa.eu/eurostat/fr/web/products-eurostat-news/-/DDN-20200603-1. Accessed 10 Jan 2023

Eurostat.a2022. *Organic Farming Statistics,* https://ec.europa.eu/eurostat/statistics-explained/index.php?title=Organic_farming_statistics#Total_organic_area [19.02.2023]

Faller A.L.K., Fialho E. Polyphenol content and antioxidant capacity in organic and conventional plant foods. *J Food Compost Anal.* 2010, 23, 6, 561–568.

FAO. FAOSTAT: Pesticide indicators-agri-environmental indicator on the use of pesticides per area of cropland; *Food and Agricultural Organisation*, Rome, Italy, 2021.

Fehér A., Gazdecki M., Véha M., Szakály M., Szakály Z. A comprehensive review of the benefits of and the barriers to the switch to a plant-based diet. *Sustainability* 2020, 12, 4136. https://doi.org/10.3390/su12104136

Fenibo E.O., Ijoma G.N., Matambo T. Biopesticides in sustainable agriculture: A critical sustainable development driver governed by green chemistry principles. *Front. Sustain. Food Syst.* 2021, 5, 619058. doi: 10.3389/fsufs.2021.619058

Fetting, C. 2020. *The European Green Deal*. Vienna: Austria.

Food Standards Agency. *McCance and Widdowson's the composition of foods*. 6th summary edition. Cambridge, United Kingdom, Royal Society of Chemistry 2002.

Francis, A. Organic farming. *In: Encyclopedia of Soils in the Environment*, D. Hillel, J.L. Hatfield (eds.), Elsevier/Academic Press, 2004.

Franco M.F.S., Delgado E.U.A. Relation of fertilization and the quality of agricultural products. *Research, Society and Development* 2022, 11, 4, e36311427562. https://doi.org/10.33448/rsd-v11i4.27562.

Fratini F., Cilia G., Mancini S., Felicioli A. Royal Jelly: An ancient remedy with remarkable antibacterial properties, *Microbiol. Res.* 2016a, 192, 130–141. https://doi.org/10.1016/j.micres.2016.06.007

Fratini F., Cilia G., Turchi B., Felicioli A. Beeswax: A mini review of its antimicrobial activity and its application in medicine. *Asian Pac. J. Trop. Med.* 2016b, Volume 9, Issue 9, 839–843. https://doi.org/10.1016/j.apjtm.2016.07.003.

Freitas S., Shinohara L., Sforcin J., Guimaraes S. In vitro effects of propolis on giardia duodenalis thophozoites. *Phytomedicine* 2006, 13, 3, 170–5. doi: 10.1016/j.phymed.2004.07.008.

Fukazawa T., Kobayashi T., Tokairin S., Chimi K., Maruyama T., Yanagita T. Behavior of N-methylcarbamate pesticides during refinement processing of edible oils. *J. Oleo Sci.* 2007, 56, 65–71. doi: 10.5650/jos.56.65.

Fulton, L., Howes, T., Hardy, J. 2004. *Biofuels for Transport: an International Perspective*. Paris: International Energy Agency.

Galani J.H.Y., Patel J.S., Patel N.J., Talati J.G. Storage of fruits and vegetables in refrigerator increases their phenolic acids but decreases the total phenolics, anthocyanins and vitamin C with subsequent loss of their antioxidant capacity. *Antioxidants* 2017, 6(3), 59. https://doi.org/10.3390/antiox6030059

Gates, B. 2021. How to avoid a climate disaster. *The Solutions We Have and Breakthroughts We Need*. New York – Penguin Random House.

Gawęcki J., Roszkowski W. *Żywienie Człowieka a Zdrowie Publiczne, cz. III. [Human Nutrition and Public Health, Part III.]*. Wyd. Nauk. PWN, Warszawa, 2009. [in Polish]

Giller K.E., Andersson J.A., Sumberg J. Thompson. A golden age for agronomy? J. Sumberg (Ed.), *Agronomy for Development: the Politics of Knowledge in Agricultural Research*, Routledge, London 2017, 150–160.

Giusti A., Pirard C., Charlier C., Petit J.C.J., Crevecoeur S., Remy S. (2018). Selection and ranking method for currently used pesticides (CUPs) monitoring in ambient air. *Air Quality, Atmosphere & Health 2018, 1–12. doi:10.1007/s11869-017-0516-6*

Gizaw Z. Public health risks related to food safety issues in the food market: a systematic literature review. *Environ. Health Prev. Med.* 2019, 24(1), 68. doi: 10.1186/s12199-019-0825-5.

Glavic P., Pintaric Z. N., Bogataj M. Process design and sustainable development—a European perspective. *Processes* 2021, 9, 148. doi: 10.3390/pr9010148

Gliński Z., Buczek K., Marć M. *Zjawiska I Mechanizmy Odporności Przeciwzakaźnej Pszczoły Miodnej – Nowe Osiągnięcia. [Phenomena and Mechanisms of Honeybee Anti-Infectious Immunity – New Achievements]*. Życie Weterynaryjne 2011, 86, 9, 687–694. [in Polish]

Godlewska M., Świsłocka R.: Fizykochemiczne i przeciwdrobnoustrojowe właściwości miodów z rejonu Podlasia [The physicochemical and antimicrobial properties of honeyfrom the region of Podlasie]. *Kosmos – Problemy Nauk Biologicznych* 2015, 64, 2, 307, 347–352. [in Polish]

Goetzberger, A., Zastrow, A. 2007. On the Coexistence of Solar-Energy Conversion and Plant Cultivation. *International Journal of Solar Energy* 1(1): 55–69.

Golijan J., Sečanski M. Organic plant products are of more improved chemical composition than conventional ones. *Food Feed Res.* 2021, 48(2), 79–117. https://doi.org/10.5937/ffr48-30907

GRAAGG. 2022. *Global Research Alliance on Agricultural Greenhouse Gases*, https://globalresearchalliance.org/ [19.02.2023]

Grewal A.S., Singla A., Kamboj P., Dua J.S. Pesticide residues in food grains, vegetables and fruits: A hazard to human health. *J. Med. Chem.* Toxicol. *2017, 2, 40–46. DOI: 10.15436/2575-808X.17.1355*

Grossman, M.R. Good agricultural practice in the united states: conservation and climate. *Environmental Law Review* 2011, 13(4), 296–317. https://doi.org/10.1350/enlr.2011.13.4.296

Gruchelski M., Niemczyk J. Substancje dodatkowe w żywności: efektywność rynku żywnościowego versus zdrowie konsumentów. [Additives in food: efficiency of the food market versus consumer health]. *Postępy Techniki Przetwórstwa Spożywczego* 2019, 1, 108–112. [in Polish]

Gulbicka B., Kwasek M. The impact of income on food consumption – premises for food policy. *Zagadnienia Ekonomiki Rolnej* 2006, 1, 19–33. https://depot.ceon.pl/handle/123456789/5610

Guo J., Wang Z., Chen Y., Cao J., Tian W., Ma B., Dong Y. Active components and biological functions of royal jelly. *J. Func. Foods* 2021, 82, 104514. https://doi.org/10.1016/j.jff.2021.104514.

Gupta A. Pesticide residue in food commodities. *Agrobios*

Guterres, A. 2020. *Carbon neutrality by 2050: The World's Most Urgent Mission*. New York: UN Secretary General.

Haiying W., Shaozhi Z., Guangming C. Experimental study on the freezing characteristics of four kinds of vegetables. *LWT-Food Science and Techn.* 2007, 40.

Hallmann E., Rembiałkowska E. Selected nutrient content in red onions from organic and conventional production. *Żywność. Nauka. Technologia. Jakość.* 2007, 2 (51), 105–111.

Harasim E., Kwiatkowski C.A. Selected offers of rural areas and sustainable agriculture and horticulture. [Wybrane oferty obszarów wiejskich oraz zrównoważonego rolnictwa i ogrodnictwa]. *Instytut Naukowo-Wydawniczy Spatium*, Radom, 2020, pp. 142, e-ISBN: 978-83-66550-01-8 [in Polish]

Harecia J.F., Garcia-Galavis P.A., Ruiz Dorado J.A., Maqueda C. Comparison of nutritional quality of the crops grown in an organic and conventional fertilized soil. *Scientia Horticulturae 2011, 129, 882–888.*

Harwood R.R., 2020. A history of sustainable agriculture. In *Sustainable agricultural systems* (3–19). CRC Press

Hedlund J., Longo S., York R. Agriculture, pesticide use, and economic development: a global examination (1990–2014). *Rural Sociol.* 2019, 85, 519–544. https://doi.org/10.1111/ruso.12303

Helfrich L.A. Aquatic animals; A Guide to Reducing Impacts on Aquatic Systems. 2009. http//publc.ext.vt.edu/420/420-013/420-013htm. Accessed on June 20, 2023.

Hellner M., Winter D., Georgi R., Munstedt K. Usage and experience in German beekeepers. *Evidence-Based Complementary and Alternative Medicine*, 2007, 5, 4, 475–479.

Henry C. Functional foods. *Eur. J. Clin. Nutr.* 2010, 64, 657–659. https://doi.org/10.1038/ejcn.2010.101

Hoang V. Modern short food supply chain, good agricultural practices, and sustainability: A conceptual framework and case study in Vietnam. *Agronomy* 2021, 11, 2408. https://doi.org/10.3390/agronomy11122408

Hock W., *Toxicity of Pesticides*, 2022. Available on-line: https://extension.psu.edu/toxicity-of-pesticides (accessed on 20 June 2023)

Hołderna-Kędzia E., Kędzia B. Research on an antioxidant capacity of honeys. *Acta Agrobot.* 2006, 59, 1, 265–269.
https://doi.org/10.3389/fsufs.2022.892818

Hurtado-Barroso S., Tresserra-Rimbau A., Vallverdú-Queralt A., Lamuela-Raventós R. M. Organic food and the impact on human health. *Critical Reviews in Food Science and Nutrition,* 2019, 59,4, 104–714. DOI: 10.1080/10408398.2017.1394815

Hussain M.A., Gooneratne R. Understanding the fresh produce safety challenges. *Foods* 2017, 21, 6(3), 23. doi: 10.3390/foods6030023.

Iberdola.com. 2020. *Kiribati, The First Country Rising Sea Levels will Swallow Up as a Result of Climate Change,* https://www.iberdrola.com/environment/kiribati-climate-change [1.10.2020].

Idah P. A., Musa J. J., Abdullahi M. Effects of Storage Period on Some Nutritional Properties of Orange and Tomato. *AU Journal of Technology* 2010, 13(3), 181–185.

IFOAM. 2022. *IFOAM Family Standards,* https://www.ifoam.bio/our-work/how/standards-certi-fication/organic-guarantee-system/ifoam-family-standards [19.02.2023].

Iguacel I., Huybrechts I., Moreno L. A., Michels N. Vegetarianism and veganism compared with mental health and cognitive outcomes: a systematic review and meta-analysis. *Nutrition Rev.* 2020, 79(4), 361–381. DOI: 10.1093/nutrit/nuaa030

Iizuka T., Shimizu A., Removal of pesticide residue from Brussels sprouts by hydrostatic pressure, *Innovative Food Science & Emerging Technologies* 2014, 22, 70–75, https://doi.org/10.1016/j.ifset.2014.01.009.

Ikerd, J. 2012. *The Essentials of Enonomic Sustainability.* Sterling: Kumarin Press.

IRENA. 2020. *Irena Renewable Energy Statistics* 2020, https://www.irena.org/-/media/Files/IRENA/Agency/Publication/2020/Jul/IRENA_Renewable_Energy_Statistics_2020.pdf [1.11.2020].

Jackson H. B, Fahrig L. Are ecologists conducting research at the optimal scale? *Glob. Ecol. Biogeogr.* 2015, 24, 52–63. https://doi.org/10.1111/geb.12233

Jacquet F., Jeuffroy MH., Jouan J. *et al.* Pesticide-free agriculture as a new paradigm for research. *Agron. Sustain. Dev.* 2022, 42, 8. https://doi.org/10.1007/s13593-021-00742-8

Jaggard, K., Qi, A. and Ober E. 2010. Possible changes to arable crop yields. *Philosophical Transactions of the Royal Society B* 365: 2835–2851.

Janicki A. Żywnosc wygodna – definicja i etapy rozwoju (Convenience food – definition and stages of development). *Przemysł Spożywczy* 1993, 47, 09, 227–230. [in Polish]

Jasińska B., Tomaka K., Uram-Dudek A., Paradowska K. Fizykochemiczna analiza miodów z rejonu Podkarpacia (Physicochemical analysis of honeys from the Podkarpacie region). *Post Fitoter* 2020, 21(4): 219–227 DOI: https://doi.org/10.25121/PF.2020.21.4.219 [in Polish]

Jones J.W., Antle J.M., Basso B., Boote K.J., Conant R.T., Foster I., Godfray H.C.J., Herrero M., Howitt R.E,, Janssen S., Keating B.A., Munoz-Carpena R., Porter C.H., Rosenzweig C., Wheeler T.R. Brief history of agricultural systems modeling. *Agric. Syst.* 2017, 155, 240–254. doi: 10.1016/j.agsy.2016.05.014.

Kakar K., Nitta Y., Asagi N., Komatsuzaki M., Shiotsu F., Kokubo T., Xuan T.D. Morphological analysis on comparison of organic and chemical fertilizers on grain quality of rice at different planting densities. *Plant Prod. Sci.* 2019, 22(4), 510–518. DOI: 10.1080/1343943X.2019.1657777

Kapusta-Duch J., Wisła-Świder A., Nowak E. Content evaluation of nitrates and nitrites in white sauerkraut kept in cold storage and produced from conventional and organic cultivation. *Żywność. Nauka. Technologia. Jakość.* 2021, 28, 2 (127), 106–20. DOI: 10.15193/zntj/2021/127/381

Kassam A. T. Friedrich, R. Derpsch. Global spread of conservation agriculture. *Int. J. Environ. Stud.* 2019, 76 (1), 29–51. DOI: 10.1080/00207233.2018.1494927

Katan M. B. Nitrate in foods: harmful or healthy? *Am. J. Clin. Nutr.* 2009, 90 (1). 11–2. DOI: 10.3945/ajcn.2009.28014

Kazimierczak R., Hallmann E., Sułek M. Content of carotenoids in selected carrot juices from organic and conventional production and intended for consumption by infants and adults. *Żywność. Nauka. Technologia. Jakość.* 2018, 25, 2 (115), 81–92. DOI: 10.15193/ZNTJ/2018/115/235

Keikotlhaile B., Spanoghe P., Steurbaut W. Effects of food processing on pesticide residues in fruits and vegetables: A metaanalysis approach. *Food Chem. Toxicol.* 2010, 48(1), 1–6. doi: 10.1016/j.fct.2009.10.031.

Khalifa S.A.M., Elashal M.H., Yosri N., Du M., Musharraf S.G., Nahar L., Sarker S.D., Guo Z., Cao W., Zou X., Abd El-Wahed A.A., Xiao J., Omar H.A., Hegazy M.F., El-Seedi H.R. Bee pollen: Current status and therapeutic potential. *Nutrients* 2021, 13(6), 1876. doi: 10.3390/nu13061876.

Khalifa, S.A.M., Elashal, M., Kieliszek, M., Ghazala, N.E., Farag, M.A., Saeed, A., Xiao, J., Zou, X., Khatib, A., Göransson, U., El-Seedi, H.R. Recent insights into chemical and pharmacological studies of bee bread, *Trends in Food Science & Technology* 2020, 97, 300–316. https://doi.org/10.1016/j.tifs.2019.08.021.

Kharel M., Dahal B.M., Raut N. Good agriculture practices for safe food and sustainable agriculture in Nepal: A review. *J. Agric. Food Res.* 2022, 10, 100447. https://doi.org/10.1016/j.jafr.2022.100447

Kiely L.J., Hickey R.M. Characterization and Analysis of Food-Sourced Carbohydrates. *Methods Mol. Biol. 2022, 2370, 67–95. doi: 10.1007/978-1-0716-1685-7_4.*

Kılıç O., Boz İ., Eryılmaz G.A. Comparison of conventional and good agricultural practices farms: A socio-economic and technical perspective. *J. Clean. Prod.* 2020, 258, 120666. https://doi.org/10.1016/j.jclepro.2020.120666.

Kim S.D., Kim I.D., Park M.Z., Lee Y.G. Effect of ozone water on pesticide-residual contents of soybean sprouts during cultivation. *Korean J. Food Sci. Technol.* 2000, 32, 277–283.

Kocot J., Kiełczykowska M., Luchowska-Kocot D., Kurzepa J., Musik I. Antioxidant potential of propolis, bee pollen, and royal jelly: Possible medical application. *Oxid Med Cell Longev.* 2018, 7074209. doi: 10.1155/2018/7074209.

Kolbe H., Stephan-beckmann S. Development, growth and chemical composition of the potato crop (solanum tuberosum L.). II. Tuber and whole plant. *Potato Res.* 1997, 40, 135–153. https://doi.org/10.1007/BF02358240

Komosinska-Vassev K., Olczyk P., Kaźmierczak J., Mencner L., Olczyk K. Bee pollen: chemical composition and therapeutic application. *Evid Based Complement Alternat Med.* 2015, 297425. doi: 10.1155/2015/297425.

Kordylas, G. 2022. Uciec, ale dokąd? *Przegląd* 7: 32–34 (in Polish)

Kortetmäki T., Oksanen M. Is there a convincing case for climate veganism? *Agric Human Values.* 2020, 38(1), 1–12. DOI: 10.1007/s10460-020-10182-x

Kotynia Z., Szewczyk P. Nadzór nad stosowaniem dodatków do żywności – bezpieczeństwo spożycia produktów przetworzonych. [Supervision over the use of food additives – safety of consumption of processed products]. *Kontrola Państwowa* 2019, 64, 3(386), 58–64. [in Polish]

Kowalczuk I. Innowacje żywności w opinii konsumentów i producentów. [Food innovations in the opinion of consumers and producers]. *Wyd. SGGW*, Warszawa, 2011. [in Polish]

Kozak L., Michałowski A., Proch J., Krueger M., Munteanu O., Niedzielski P. Iron forms Fe(II) and Fe(III) Determination in pre-roman iron age archaeological pottery as a new tool in archaeometry. *Molecules* 2021, 26, 5617. https://doi.org/10.3390/molecules26185617

Kramkowska M., Grzelak T., Czyżewska K. Benefits and risks associated with genetically modified food products. *Ann. Agric. Environ. Med.* 2013, 20(3), 413–419. PMID: 24069841

Krasilnikov P., Taboada M.A., Amanullah. Fertilizer Use, Soil Health and Agricultural Sustainability. *Agriculture* 2022, 12, 462. https://doi.org/10.3390/agriculture12040462

Kremen C., Miles A. Ecosystem services in biologically diversified versus conventional farming systems: Benefits, externalities, and trade-offs. *Ecol. Soc.* 2012, 17(4), 40. http://dx.doi.org/10.5751/ES-05035-170440

Kubisz P., Dalton G., Majewski E., Pogodzińska K. Facts and myths about GM food—the case of poland. *Agriculture* 2021, 11, 791. https://doi.org/10.3390/agriculture11080791

Kuepper G., Gegner L. Organic crop production. Overview. *NCAT Agriculture Specialists* 2004.

Kumar, P. 2017. *Largest biogas plants*, https://www.nsenergybusiness.com/news/newslargest-biogas-plants-061017-5943061/# [1.11.2020].

Kwasek M., Obiedzińska A. *Z Badań Nad Rolnictwem Społecznie Zrównoważonym, Zrównoważone Systemy Rolnicze I Zrównoważona Dieta. Nr 119, Seria Program Wieloletni 2011–2014. [From Research on Socially Sustainable Agriculture, Sustainable Agricultural Systems and a Balanced Diet. No. 119, series Multiannual Program 2011–2014]*. IERiGŻ-PIB, Warszawa, 2014. [in Polish]

Kwiatkowski C., 2022, *Unpublished Data*.

Kwiatkowski C.A., Haliniarz M., Kołodziej B., Harasim E., Tomczyńska-Mleko M. Content of some chemical components in carrot (*Daucus carota* L.) roots depending on growth stimulators and stubble crops. *J. Elem.* 2015, 20(4), 933–943. DOI: 10.5601/jelem.2014.19.4.812

Kwiatkowski C.A., Harasim E. Produkcja rolnicza a bezpieczna żywność – wybrane aspekty. [Agricultural production and safe food – selected aspects]. *Instytut Naukowo-Wydawniczy Spatium*, Radom, 2019, pp. 146. ISBN: 978-83-66017-61-0 [in Polish]

Lairon D. Nutritional quality and safety of organic food. A review. *Agron Sustain Dev.* 2010, 30 (1), pp. 33–41. ff10.1051/agro/2009019ff. ffhal-00886513f

Lee R., den Uyl R., Runhaar H. Assessment of policy instruments for pesticide use reduction in Europe; learning from a systematic literature review. *Crop Prot.* 2019, 126, 104929. https://doi.org/10.1016/j.cropro.2019.104929

Lennon S.F., Reighard G.L., Horton D., Schermerhom P., Podhorniak L., Infante R. Profiling presence and concentration of eighteen pesticide residues through a commercial canning process. *Acta Hort.* 2006, 713, 409–415.

Leventon J., Laudan J. Local food sovereignty for global food security? Highlighting interplay challenges. *Geoforum* 2017, 85, 23–26. http://dx.doi.org/10.1016/j.geoforum.2017.07.002

Li M., Liu S., Sun Y., Liu Y. Agriculture and animal husbandry increased carbon footprint on the Qinghai-Tibet Plateau during past three decades. *J. Clean. Prod.* 2021, 278, 123963. https://doi.org/10.1016/j.jclepro.2020.123963.

Li Y.C., Li Z.W., Lin W.W., Jiang Y.H., Weng B.Q., Lin W.X. Effects of biochar and sheep manure on rhizospheric soil microbial community in continuous ratooning tea orchards. *Chin. J. Appl. Ecol.* 2018, 29, 1273–1282. doi: 10.13287/j.1001-9332.201804.036.

Lin B., Luo G.H., Xu Q.X., Wang Q.S., Guan X.F. Effects of biogas residue on yield and quality of tea. *Fujian J. Agric. Sci.* 2010, 25, 90–95.

Lin W., Lin M., Zhou H., Wu H., Li Z., Lin W. The effects of chemical and organic fertilizer usage on rhizosphere soil in tea orchards. *PLoS One.* 2019, 14(5), e0217018. doi: 10.1371/journal.pone.0217018.

Lisboa C.F., de Castro Melo E., Lopez Donzeles S.M. *Influence of Storage Conditions on Quality Attributes of Medicinal Plants. Biomedical Journal of Scientific & Technical Research (BJSTR)*, 2018, 4, 4093–4095. DOI: 10.26717/BJSTR.2018.04.001097.

Lotz L.A.P., van de Wiel C.C.M., Smulders M.J.M. How to assure that farmers apply new technology according to good agricultural practice: Lessons from Dutch initiatives. *Front. Environ. Sci.* 2018, 6, 89. doi: 10.3389/fenvs.2018.00089

Łukasiewicz A. The use of bee venom in cosmetology. *Aesthetic Cosmetology and Medicine* 2021, 10,1, 23–31.

Luo H., Robles-Aguilar A.A., Sigurnjak I., Michels E., Meers E. Assessing nitrogen availability in biobased fertilizers: Effect of vegetation on mineralization patterns. *Agriculture* 2021, 11, 870. https://doi.org/10.3390/agriculture11090870

Maćkow, J., Paczosa, A. & Skirmuntt, G. 2004. *Eko-Generacja Przyszłosci*. Katowice, Warsaw: WNS.

Mæhre H.K., Dalheim L., Edvinsen G.K., Elvevoll E.O., Jensen I.J. Protein Determination—Method Matters. *Foods* 2018, 7(1), 5. DOI: 10.3390/foods7010005

Maghoumi M., Amodio M.L., Fatchurrahman D., Cisneros-Zevallos L., Colelli G. Pomegranate husk scald browning during storage: A review on factors involved, their modes of action, and its association to postharvest treatments. *Foods* 2022, 11, 3365. https://doi.org/10.3390/foods11213365

Maidana N.L., Vanin V.R., Horii C.L., Ferreira F.A., Rajput M.U. Absolute determination of soluble potassium in tea infusion by gamma-ray spectroscopy. *Food Chem.* 2009, 116, 2, 555–560. https://doi.org/10.1016/j.foodchem.2009.03.005.

Majewska E., Drużyńska B., Derewiaka D., Ciecierska M., Wołosiak R.: Fizykochemiczne wyróżniki jakości wybranych miodów nektarowych [Physicochemical parameters of quality of floral honey] *Bromat. Chem. Toksykol.* 2015, 48, 3, 440–444. [in Polish]

Majkowska-Gadomska J., Mikulewicz E., Francke A. Effects of plant covers and mulching on the biometric parameters, yield and nutritional value of tomatillos (*Physalis ixocarpa* Brot. Ex Hornem.). *Agronomy* 2021, 11, 1742. https://doi.org/10.3390/agronomy11091742

Mamzer H. *Wybryk Natury czy Natura? Socjologiczne Uwarunkowania Wzrostu Popularności Diet Bezmięsnych. [Freak of Nature or Nature? Sociological Determinants of the Increase in the Popularity of Meat-free Diets].* Wydawnictwo Naukowe Wyższej Szkoły Biznesu i Nauk o Zdrowiu Łódź, 2020. https://www.academia.edu/377200 [in Polish]

Mărgăoan R., Stranț M., Varadi A., Topal E., Yücel B., Cornea-Cipcigan M., Campos M.G., Vodnar D.C. Bee collected pollen and bee bread: Bioactive constituents and health benefits. *Antioxidants* 2019, 8(12), 568. doi: 10.3390/antiox8120568.

Markuszewski B., Kopytowski J. Evaluation of plum cultivars grafted on 'Wangenheim Prune' rootstock in the northeast of Poland. *Folia Hort.* 2013, 25, 2, 101–106. https://doi.org/10.2478/fhort-2013-0011

Marrone P.G. Pesticidal natural products–status and future potential. *Pest. Manag. Sci.* 2019, 75, 2325–2340. doi: 10.1002/ps.5433

Masaba K.G., Nguyen M. Determination and comparison of vitamin C, calcium and potassium in four selected conventionally and organically grown fruits and vegetables. *Afr. J. Biotechnol.* 2008, 7(16), 2915–2919. http://www.academicjournals.org/AJB

Matysiak J., Matysiak J. Kokot Z.J.: Właściwości farmakologiczne jadu pszczelego [Pharmacological properties of honeybee venom]. *Nowiny Lekarskie* 2008, 77, 6, 435–458. [in Polish]

Maul J.D., Blackstock C., Brain R.A. Derivation of avian dermal LD50 values for dermal exposure models using in vitro percutaneous absorption of [14C]-atrazine through rat, mallard, and northern bobwhite full thickness skin. *Sci Total Environ.* 2018, 630, 517–525. doi:10.1016/J.SCITOTENV.2018.02.206

Meadows, D.H., Meadows D.L., Randers, J. 1992. Beyond the limits. *Global Collapse or a Sustainable Future.* London: Earthscan Publications Limited.

Mie A., Andersen H.R., Gunnarsson S., Kahl J., Kesse-Guyot E., Rembiałkowska E., Quaglio G., Grandjean P. Human health implications of organic food and organic agriculture: a comprehensive review. *Environ Health.* 2017, 16(1), 111. doi:10.1186/s12940-017-0315-4.

Millennium Ecosystem Assessment. 2005. *Living Beyond Our Means. Natural Assests and Human Well-Being. Statement From The Board.* New York: United Nationss Environmental Programme.

Mineau, P. 2022. Neonic Insecticides and invertebrate species endangerment. In: Della Sala D.A. and Goldstein M.I. (eds.), *Imperiled: The Encyclopedia of Conservation.* Amsterdam: Elsevier.

Miras-Avalos, J.M., Baveye P.C. 2018. Agroecosystems facing global climate change: The search for sustainability, *Frontiers in Environmental Science* 6: 608.

Moragues-Faus, A. Problematising justice definitions in public food security debates: towards global and participative foodjustices. *Geoforum* 2017, 84, 95–106. http://dx.doi.org/10.1016/j.geoforum.2017.06.007.

Mukherjee A., Omondi, E.C., Hepperly P.R., Seidel R., Heller W.P. Impacts of organic and conventional management on the nutritional level of vegetables. *Sustainability* 2020, 12, 8965. https://doi.org/10.3390/su12218965

Muller A., Schader C., El-Hage Scialabba N. *et al.* Strategies for feeding the world more sustainably with organic agriculture. *Nat. Commun.* 2017, 8, 1290. https://doi.org/10.1038/s41467-017-01410-w

Murawska B., Piekut A., Jachymska J., Mitura K., Lipińska K.J. Ekologiczny i konwencjonalny system gospodarowania a wielkość i jakość plonu wybranych roślin uprawnych. [Ecological and conventional farming system vs. quantity and quality of yield of selected cultivated plants]. *Infrastruktura i Ekologia Terenów Wiejskich* 2015, 3(1), 663–675. [in Polish]

Nachtman G. Farms combining organic and conventional production methods at the background of organic farms. *Zagadnienia Ekonomiki Rolnej* 2015, 3(344), 129–147. DOI: 10.5604/00441600.1167241

Nainu F., Masyita A., Bahar A. M., Raihan M., Prova S.R., Mitra S., Emran T.B., Simal-Gandara J. Pharmaceutical prospects of Bee products: Special focus on anticancer, antibacterial, antiviral, and antiparasitic properties. *Antibiotics* 2021, 10, 7, 822. https://doi.org/10.3390/antibiotics10070822

Navarro S., Perez G., Navarro G., Vela N. Decline of pesticide residues from barley to malt. *Food Addit. Cont.* 2007, 24, 851–859. DOI: 10.1080/02652030701245189

Neff R.A., Hartle J.C., Laestadius L., Dolan K., Rosenthal A.C., Nachman K.E. A comparative study of allowable pesticide residue levels on produce in the United States. *Glob. Health* 2012, 8, 2. DOI:10.1186/1744-8603-8-2

Nicolopoulou-Stamati P., Maipas S., Kotampasi C., Stamatis P., Hens L. Chemical pesticides and human health: The urgent need for a new concept in agriculture. *Front. Public Health* 2016, 4:148. doi: 10.3389/fpubh.2016.00148

Nielsen S.S. *Food Analysis.* Fourth edition. Springer, New York, 2010.

Obey J.K., Ngeiywa M.M., Lehesvaara, M. *et al.* Antimicrobial activity of commercial organic honeys against clinical isolates of human pathogenic bacteria. *Org. Agr.* 2022, 12, 267–277. https://doi.org/10.1007/s13165-022-00389-z

Obiedzińska A., Waszkiewicz-Robak B. Oleje tłoczone na zimno jako żywność funkcjonalna. *Żywność. Nauka. Technologia. Jakość.* 2012, 1(80), 27–44.

Official Journal of the European Communities, 1999 (http://eur-lex.europa.eu/LexUriServ/LexUriServ.do?uri=OJ:L:1999:222:0001:0028:EN:PDF

Orden L., Ferreiro N., Satti P., Navas-Gracia L.M., Chico-Santamarta L., Rodríguez R.A. Effects of onion residue, bovine manure compost and compost tea on soils and on the agroecological production of onions. *Agriculture* 2021, 11, 962. https://doi.org/10.3390/agriculture11100962

Organic Production Logo https://ec.europa.eu/info/food-farming-fisheries/farming/organic-farming/organic-logo_pl (access 07.02.2023)

Ostandie N., Giffard B., Bonnard O. *et al.* Multi-community effects of organic and conventional farming practices in Vineyards. *Sci. Rep.* 2021, 11, 11979. https://doi.org/10.1038/s41598-021-91095-5

Ötles S., Ozgoz S. Health effects of dietary fiber. *Acta Sci. Pol. Technologia Alimentaria* 2014, 13 (2), 191–202. DOI:10.17306/J.AFS.2014.2.8

Oueriemmi H., Kidd P.S., Trasar-Cepeda C., Rodríguez-Garrido B., Zoghlami R.I., Ardhaoui K., Prieto-Fernández Á., Moussa M. Evaluation of composted organic wastes and farmyard manure for improving fertility of poor sandy soils in arid regions. *Agriculture* 2021, 11, 415. https://doi.org/10.3390/agriculture11050415

Pacific Institute. 2022. *California Drought, Conditions and Impact.* California Drought Org., http://www.californiadrought.org/drought/background/ [15.12.2022].

Pandey A., Hofer R., Taherzadeh M., Nampoothiri K.M., Larroche C., 2015. *Industrial Biorefineries & White Technology*, Amsterdam: Elsevier.

Pandey R.M., Upadhyay S.K. *Food Additive*. IntechOpen 2012. DOI: 10.5772/34455

Paradowska K., Zielińska A., Krawiec N. Composition and antioxidant properties of colour fraction isolated from the bee pollen. *Post. Fitoterapii*, 2014, 4, 209–215.

Pasupuleti V.R., Sammugam L., Ramesh N., Gan S.H. Honey, Propolis, and Royal Jelly: A comprehensive review of their biological actions and health benefits. *Oxid Med Cell Longev.* 2017, 1259510. doi: 10.1155/2017/1259510.

Patinha C., Durães N., Dias A.C., Pato P., Fonseca R., Janeiro A. *et al.* Long-term application of the organic and inorganic pesticides in vineyards: Environmental record of past use. *Appl. Geochem.* 2018, 88, 226–238. doi:10.1016/J.APGEOCHEM.2017.05.014

Paulino N., Coutinho L.A., Coutinho J.R., Vilela G.C., da Silva Leandro V.P., Paulino A.S. Antiulcerogenic effect of Brazilian propolis formulation in mice. *Pharmacology & Pharmacy* 2015, 6, 12. DOI: 10.4236/pp.2015.612060

Paull J., Attending the first organic agriculture course: Rudolf Steiner's agriculture course at Koberwitz, 1924, *European Journal of Social Sciences* 2011, 21 (1), 64–70

Pawłowski A. 2020. Sustainable energy: Technology, industry, transport and agriculture. In: Pawłowski L., Litwińczuk Z., Zhou G. (eds), *The Role of Agriculture in Climate Change Mitigation*. Boca Raton, London, New York: CRC Press, Balkema, Taylor & Francis Group: 1–17.

Pawłowski, A. 2011. *Sustainable Development as a Civilizational Revolution. Multidimensional Approach to the Challenges of the 21st century*. Boca Raton, Londyn, New York, Leiden: CRC Press, Taylor & Francis Group, A Balkema Book.

Pawłowski, A. Sustainable development and renewable sources of energy. In: *Advances in Environmental Engineering Research in Poland*, Pawłowska, M., Pawłowski, L. (eds), Routledge: London, Great Britain, 2021; pp. 3–15.

Pereira E.M.R., da Silva J.L.D.C., Silva F.F., De Luca M.P., Lorentz T.C.M., Santos V.R. Clinical evidence of the efficacy of a mouthwash containing propolis for the control of plaque and gingivitis: A phase II study. *Evidence-Based Complementary and Alternative Medicine* 2011, 750249. doi: 10.1155/2011/750249. 2011.

Perez-Arqillue C., Conchello P., Arino A., Juan T., Herrera A. Quality evaluation of Spanish rosemary (*Rosmarinus Officinalis*) honey. *Food Chem.* 1994, 51, 207–210.

PET. 2022. *Factory Farming – Misery for Animals*, https://www.peta.org/issues/animals-used-for-food/factory-farming/ [19.02.2023].

Pig Progress. 2018. *ASF update: Outbreaks in Romania, EU Steps up the Fight*, https://www.pigprogress.net/health-nutrition/asf-update-outbreaks-in-romania-eu-steps-up-the-fight/ [19.02.2023].

PN-90/A-75101/03 *Przetwory Owocowe i Warzywne. Oznaczanie Zawartości Suchej Masy Metodą Wagową. [Fruit and Vegetable Preserves. Determination of Dry Matter Content by Gravimetric Method]*. Warszawa: PKN, 1990. [in Polish]

PN-A-79011-15:1998 *Koncentraty Spożywcze. Oznaczanie Zawartości Błonnika Pokarmowego. Przetwory Owocowe i Warzywne. [Food Concentrates. Determination of Dietary Fiber Content. Fruit and Vegetable Preserves]*. Warszawa, PKN, 1998. [in Polish]

Polak-Śliwińska M., Tańska M. Conventional and organic honeys as a source of water- and ethanol-soluble molecules with nutritional and antioxidant characteristics. *Molecules* 2021, 26, 3746. https://doi.org/10.3390/molecules26123746

Ponting, C. 1993. *A Green History of the World: The Environment and the Collapse of Great Civilizations*. New York: Penguin Books.

Popa M.E., Mitelut A.C., Popa E.E. Vlad A.S., Popa I. Organic foods contribution to nutritional quality and value. *Trends. Food Sci. Tech.* 2019, 84, 15–18. https://doi.org/10.1016/j.tifs.2018.01.003

Popov-Raljić J., Arsić N., Zlatković B., Basarin B., Mladenović M., Lalicić-Petronijević J., Ivkov M., Popov V. Evaluation of color, mineral substances and sensory uniqueness of meadow and acacia honey from Serbia. *Rom. Biotechnol. Lett.* 2015, 20, 10784–10799.

Popp, J., Lakner, Z., Harangi-Rakos, M., Fari, M. 2014. The effect of bioenergy expansion: Good, energy and environment. *Renewable and Sustainable Energy Reviews* 32: 559–578.

Prędka A., Gronowska-Senge A. Antioxidant properties of selected vegetables from organic and conventional system of cultivation in reducing oxidative stress. *Żywność. Nauka. Technologia. Jakość.* 2009, 4 (65), 9–18. [in Polish]

Public Citizen, *Protecting Health, Safety & Democracy*, http://www.citizen.org/cmep/foodsafety/eu/eucafo/pol/index.cfm [30.06.2009].

Publications (India), 2006, pp. 285. ISBN-13:978-8177542677

Pucca B.M., Cerni A.F., Oliveira S.I., Jenkins T.P., Argemi L., Sorensen Y.C., Ahmadi S., Barbosa E. J., Laustsen H. A. Bee updated: Current konwledge on bee venom and bee envenoming therapy. *Front. Immunol.* 2019, 10. https://doi.org/10.3389/fimmu.2019.02090

Rahman M.S., Labuza T.P. *Handbook of Food Preservation.* Third edition. CRC Press Taylor&Francis Group 2020. https://doi.org/10.1201/9780429091483

Rajput H., Rehal J. Methods for food analysis and quality control. *State of the Art Technologies in Food Science* 2019, 229–346.

Raman R. The impact of Genetically Modified (GM) crops in modern agriculture: A review. *GM Crops Food.* 2017, 8(4), 195–208. doi: 10.1080/21645698.2017.1413522.

Rani L., Thapa K., Kanojia N., Sharma N., Singh S., Singh Grewal A., Lal Srivastav A., Kaushal J. An extensive review on the consequences of chemical pesticides on human health and environment, *J. Cleaner Prod. 2021, 283, 124657. https://doi.org/10.1016/j.jclepro.2020.124657.*

Ratkovska B., Kunachowicz H., Przygoda B. Domestic market of food products fortified by vitamins and minerals in the light of the European regulations. *Żywność. Nauka. Technologia. Jakość.* 2007, 6 (55).

Regueiro J., López-Fernández O., Rial-Otero R., Cancho-Grande B., Simal-Gándara J. A review on the fermentation of foods and the residues of pesticides — biotransformation of pesticides and effects on fermentation and food quality, *Critical Reviews in Food Science and Nutrition* 2015, 55(6). DOI:10.1080/10408398.2012.677872

Regulation (EU) *2018/848 of the European Parliament and of the Council of 30 May 2018 on organic Production and Labelling of Organic Products and Repealing Council Regulation (EC) No 834/2007, 2018 (OJ L 150/32)* Available online: https://eur-lex.europa.eu/legal-content/EN/TXT/PDF/?uri=CELEX:32018R0848 [last accessed 12 March 2023].

Regulation of the Minister of Agriculture and Rural Development on detailed requirements for commercial quality of honey, *Journal of Laws* 2003 No. 181 Pos. 1772 and 1773 (in Polish).

Regulation of the Minister of Agriculture and Rural Development of 29 May 2015 amending the regulation on detailed requirements for the commercial quality of honey, *Journal of Laws* 2015 Pos. 850 [in Polish]

Regulation of the Minister of Agriculture and Rural Development of January 14, 2009 establishing methods of analysis relating to evaluation of honey, *Journal of Laws* from 2009 No. 17 item 94, (in Polish).

Rembiałkowska E. Quality of plant products from organic agriculture. *Journal of the Science of Food and Agriculture* 2007, 87, 2757–2762. https://doi.org/10.1002/jsfa.3000

Rembiałkowska E., Załęcka A., Badowski M. Ploeger A. The quality of organically produced food. In: Petr Konvalina (Eds.), *Organic Farming and Food Production* – November 2012. https://doi.org/10.5772/54525

Rochalska M., Orzeszko-Rywka A., Czapla K. The content of nutritive substances in strawberries according to cropping system. *J. Res. Appl. Agric. Eng.* 2011, 56(4).

Rodrigues N. da R., Barbosa Junior, J.L. and Barbosa, M.I.M.J. Determination of physicochemical composition, nutritional facts and technological quality of organic orange and purple-fleshed sweet potatoes and its flours. *Int. Food Res. J.* 2016, 23(5), 2071–2078.

Romeh A.A., Mekky T.M., Ramadan R.A., Hendawi M.Y. Dissipation of profenofos, imidacloprid and penconazole in tomato fruits and products. *Bull Env Cont Toxic.* 2009, 83(6), 812–817. doi:10.1007/s00128-009-9852-z.

Room, J. 2022. *Climate Change, What Everyone Needs to Know.* New York: Oxford University Press.

Rosati A., Borek R., Canali S. Agroforestry and organic agriculture. *Agrofor. Syst.* 2020, 95, 805–821, DOI:10.1007/s10457-020-00559-6

Roy N.K. Pesticide residues and their environmental implications. In: Roy NK, editor. *Chemistry of Pesticides* 2002, New Delhi: CBS, 265–279.

Rozman A.S., Hashim N., Maringgal B., Abdan K. A comprehensive review of stingless bee products: Phytochemical composition and beneficial properties of honey, propolis, and pollen. *Appl. Sci.* 2022, 12, 6370. https://doi.org/10.3390/app12136370

Ruiz-Mendez M.V., de la Rosa I.P., Jimenez-Marquez A., Uceda-Ojeda M. Elimination of pesticides in olive oil by refining using bleaching and deodorization. *Food Addit Contam.* 2005, 22, 23–30. doi: 10.1080/02652030400027946.

Rutkowska B. Nitrate and nitrite content in potatoes from ecological and conventional farms. *Roczniki Państwowego Zakładu Higieny / Annals of the National Institute of Hygiene* 2001, 52, 3, 231–236.

Sahm H., Sanders J., Nieberg H., Behrens G., Kuhnert H., Strohm R., Hamm U. Reversion from organic to conventional agriculture: A review. *Renewable Agriculture and Food Systems* 2013, 28(3), 263–275. doi:10.1017/S1742170512000117.

Sanaullah M., Usman M., Wakeel A., Cheema S.A., Ashraf I., Farooq M. Terrestrial ecosystem functioning affected by agricultural management systems: A review. *Soil Tillage Res.* 2020, 196. DOI: 10.1016/j.still.2019.104464

Sangboonruang, S.; Kitidee, K.; Chantawannakul, P.; Tragoolpua, K.; Tragoolpua, Y. Melittin from *Apis florea* Venom as a promising therapeutic agent for skin cancer treatment. *Antibiotics* 2020, 9, 517. https://doi.org/10.3390/antibiotics9080517

Sarwar M. Inorganic insecticides used in landscape settings and insect pests. *Chem. Res. J.* 2016, 1 (1), 50–57. http://chemrj.org/download/vol-1-iss-1-2016/chemrj-2016-01-01-50-57.pdf

Sassolas A., Prieto-Simón B., Marty J.L. Biosensors for pesticide detection: New trends. *Am. J. Anal. Chem.* 2012, 3(3). DOI: 10.4236/ajac.2012.33030

Sbahi A., Abdelwahed W., Sakur A.A. A new flame AAS application for magnesium determination in solid pharmaceutical preparations as an active ingredient and an excipient. *Int. Res. J. Pure Appl. Chem.* 2020, 21(23), 89–95. https://doi.org/10.9734/irjpac/2020/v21i2330305

Schirone M., Visciano P., Tofalo R., Suzzi G. Editorial: Biological hazards in food. *Front. Microbiol.* 2017, 7, 2154. doi:10.3389/fmicb.2016.02154.

Schmit T.M., Wall G.L., Newbold E.J., Bihn E.A. Assessing the costs and returns of on-farm food safety improvements: A survey of Good Agricultural Practices (GAPs) training participants. *PLoS One* 2020, 15(7), e0235507. https://doi.org/10.1371/journal.pone.0235507

Seufert V., Ramankutty N., Foley J.A. Comparing the yields of organic and conventional agriculture. *Nature* 2012, 485 (7397), 229–232. DOI: 10.1038/nature11069, 229–U113

Sharma A. Food additives from an organic chemistry perspective. *MOJ Biorg Org. Chem.* 2017, 1 (3), 00015. DOI:10.13140/RG.2.2.17321.70249

Shennan C., Krupnik T.J., Baird G., Cohen H., Forbush K.,, Lovell R.J., Olimpi E.M. Organic and conventional agriculture: A useful framing? *Annu. Rev. Environ. Resour.* 2017, 42 (1), 317–346. DOI: 10.1146/annurev-environ-110615-085750

Shi R., Irfan M., Liu G., Yang X., Su X. Analysis of the impact of livestock structure on carbon emissions of animal husbandry: A sustainable way to improving public health and green environment. *Front. Public Health.* 2022, 10, 835210. doi:10.3389/fpubh.2022.835210.

Shruthi E., Suma B.S. Health from the Hive: Potential uses of propolis in general health. *Int. J. Clinic. Med.* 2012, 3, 159–162. http://dx.doi.org/10.4236/ijcm.2012.33033

Shvachko N.A., Loskutov I.G., Semilet T.V., Popov V.S., Kovaleva O.N., Konarev A.V. Bioactive components in oat and barley grain as a promising breeding trend for functional food production. *Molecules* 2021, 26, 2260. https://doi.org/10.3390/molecules26082260

Sikora M., Hallmann E., Rembiałkowska E. The content of bioactive compounds in carrots from organic and conventional production in the context of health prevention. *Rocz. Państw. Zakł. Hig.* 2009, 60(3), 217–220.

Silvano M.F., Varela M.S., Palacio M.A., Ruffinengo S., Yamul D.K. Physicochemical parameters and sensory properties of honeys from Buenos Aires region. *Food Chem.* 2014, 152, 500–507. doi: 10.1016/j.foodchem.2013.12.011.

Singh P.K., Singh R.P., Singh P., Singh R.L. Chapter 2 – food hazards: Physical, chemical, and biological, Editor(s):Ram Lakhan Singh, Sukanta Mondal, *Food Safety and Human Health*, Academic Press, 2019, 15–65. https://doi.org/10.1016/B978-0-12-816333-7.00002-3.

Singleton V., Orthofer R., Lamuela-Raventós R. M. Analysis of total phenols and otheroxidation substrates and antioxidants by means of Folin-Ciocalteu reagent. *Methods Enzymol.* 1999, 299, 152–178. http://dx.doi.org/10.1016/S0076-6879(99)99017-1

Sitarz S., Janczar-Smuga M. Contemporary food safety hazards, possibilities of their control and elimination. *Nauki Inż. Technol.* 2012, 2, 5, 68–93.

Śliwińska A., Bazylak G. Evaluation of honey quality with use of selected physicochemical and microbiological parameters. *Bromat. Chem. Toksykol.* 2011, XLIV, 3, 784–791.

Śliwińska A., Przybylska A., Bazylak G. Wpływ zmian temperatury przechowywania na zawartość 5-hydroksymetylofurfuralu w odmianowych i wielokwiatowych miodach pszczelich (The influence of changes in storage temperature on the content of 5-hydroxymethylfurfural in varietal and multiflower bee honeys). *Bromat. Chem. Toksykol. – XLV, 2012, 3, str. 271–279*

Smith O.M., Cohen A.L., Rieser C.J., Davis A.G., Taylor J.M., Adesanya A.W., Jones M.S., Meier A.R., Reganold J.P., Orpet R.J., Northfield T.D., Crowder D.W. Organic farming provides reliable environmental benefits but increases variability in crop yields: A global meta-analysis. *Front. Sustain. Food Syst.* 2019, 3, 82. doi:10.3389/fsufs.2019.00082

Soares S., Amaral J.S., Oliveira M.B.P.P., Mafra I. A comprehensive review on the main honey authentication issues: Production and origin. *Compr. Rev. Food Sci. Food Saf.* 2017, 16(5), 1072–1100. Portico. https://doi.org/10.1111/1541-4337.12278

Sobol, G. 1996.Permaculture, sustainable agriculture. In: R. Jones, M. Summers & E. Mayo (eds), *Sustainable Agriculture*. London: The New Economic Foundation.

Socha R., Habryka C., Juszczak L. Effect of bee bread additive on content of phenolic compounds and antioxidant activity of honey. *Żywność. Nauka. Technologia. Jakość. 2018, 25, 2, 115, 108–119.*

Średnicka-Tober D., Obiedzińska A., Kazimierczak R., Rembiałkowska E. Environmental impact of organic vs. conventional agriculture – a review. *Journal of Research and Applications in Agricultural* 2016, 61, 204–211.

SRU (Sachverstandigenerat fur Umweltfragen). 2007. *Climate Change Mitigation by Biomass*. Executive Summary of the SRU Spécial Report, Berlin: SRU.

Ssemugabo C., Guwatudde D., Ssempebwa J.C., Bradman A. Pesticide residue trends in fruits and vegetables from farm to Fork in Kampala metropolitan area, Uganda—a mixed methods study. *Int. J. Environ. Res. Public Health* 2022, 19, 1350. https://doi.org/10.3390/ijerph19031350

Status of acute systemic toxicity testing requirements and data uses by U.S. regulatory agencies. *Regulatory Toxicology and Pharmacology* 2018, 94, 183–196. doi:10.1016/J.YRTPH.2018.01.022

Strickland J., Clippinger A.J., Brown J., Allen D., Jacobs A., Matheson J., Casey W.

Su, L., Jing, L., Zeng X., Chen, T., Liu, H., Kong, Y., Wang, X., Fu, C. and Huang, D. 2013. 3D-printed prolamin scaffolds for cell-based meat culture. *Advanced Materials* 35(2): 2207397.

Sumberg J., Giller K.E. What is 'conventional' agriculture? *Global Food Security* 2022, 32 100617. https://doi.org/10.1016/j.gfs.2022.100617

Summers, M. 1996. Overview of agriculture in Eastern Europe. In: R. Jones, M. Summers & E. Mayo (eds), *Sustainable Agriculture*. London: The New Economic Foundation.

Sun QR, Xu Y, Xiang L, Wang GS, Shen X, Chen XS, *et al.* Effects of a mixture of bacterial manureand biochar on soil environment and physiological characteristics of Mals huupehens seedlings. *Chin Agric Sci Bull.* 2017, 33, 52–59.

Svečnjak L., Chesson L.A., Gallina A., Maia M., Martinello M., Mutinelli F., Muz M.N., Nunes F.M., Saucy F., Tipple B.J., Wallner K., Waś E., Waters T.A. Standard methods for *Apis mellifera* beeswax research, *J. Apicultural Res.* 2019, 58, 2, 1–108. DOI:10.1080/00218839.2019.1571556

Świderski F. (eds.). *Żywność Wygodna i Żywność Funkcjonalna. [Convenience foods and functional foods].* WNT, Warszawa, 2003, pp. 390. ISBN: 83-204-3228-6 [in Polish]

Szeleszczuk Ł., Zielińska-Pisklak M., Goś P. Propolis – panaceum prosto z ula. [Propolis – a panacea straight from the hive]. *Farmakoterapia* 2013, 2, 23, 6–7'13 (266/26013). [in Polish]

Szwedziak K., Smolke Z., Polańczyk E., Szopa A., Koronczok J.: Metody oceny jakości miodu. [Methods for establishing honey quality]. *Postępy Techniki Przetwórstwa Spożywczego* 2017, 1, 36–38. [in Polish]

Tal A. Making conventional agriculture environmentally friendly: Moving beyond the glorification of organic agriculture and the demonization of conventional agriculture. *Sustainability* 2018, 10(4), 1078. https://doi.org/10.3390/su10041078

Temple N.J. A rational definition for functional foods: A perspective. *Front. Nutr.*, 2022, Sec. Nutrition Methodology Vol. 9. https://doi.org/10.3389/fnut.2022.957516

Teng P.S., Penning de Vries F.W.T., (editors). *Systems Approaches for Agricultural Development.* Elsevier 1992, pp. 309 (Applied Science). (ISBN: 1851668918)

Tengstrom, E. & Thynell, M. 1997. *Towards Sustainable Mobility. Transporting People and Goods in the Baltic Region.* Uppsala: Ditt Tryckeri.

Thornton, P. 2010. Livestock production: Recent trends, future prospects. *Philisophical Transactions of the Royal Society B* 365: 2853–2867.

Tomer, V., Sangha J.K. Vegetable processing at household level: Effective tool against pesticide residue exposure. *Environ. Sci. Toxicol. Food Technol.* 2013, 6, 43–53

Tomes D., Lakshmanan P., Songstad D., 2011. *Biofuels, Global Impact on Renewable Energy, Production Agriculture and Technological Advacements,* New York, Dordrecht, Heidelberg, London: Springer.

Tońska E., Toński M., Klepacka J., Łuczyńska J., Paszczyk B. The content of cadmium and lead in carrots originating from organic and conventional cultivation. *Fragm. Agron.* 2017, 34(4), 190–196.

Tosi E., Martinet R., Ortega M., Lucero H., Ré E. Honey diastase activity modified by heating, *Food Chem.* 2008, 106, 3, 883–887. https://doi.org/10.1016/j.foodchem.2007.04.025.

Trumbeckaite S., Dauksiene J., Bernatoniene J., Janulis V.: Knowledge, attitudes, and usage of apitherapy for disease prevention and treatment among undergraduate pharmacy students in Lithuania. *Evidence-Based Complementary and Alternative Medicine*, 2015, 172502. doi: 10.1155/2015/172502.

Tsvetkov I., Atanassov A., Vlahova M., Carlier L., Christov N., Lefort F., Rusanov K., Badjakov I., Dincheva I., Tchamitchian M., Rakleova G., Georgieva L., Tamm L., Iantcheva A., Herforth-Rahmé J., Paplomatas E., Atanassov I. Plant organic farming research – current status and opportunities for future development. *Biotechnol.* Biotechnol. Equipt. *2018, 32, 2, 241–260, DOI:10.1080/13102818.2018.1427509*

U.S. V. Smithfield Foods Inc. 2007. *Civil Action 1:03-CV-00434.*

UN. 2006. *Framework Convention on Climate Change.* Handbook. Bonn: Climate Change Secretariat.

UN. 2015. *Sustainable Development Goals,* https://sdgs.un.org/goals/ [19.02.2023].

UNDP. 2000. *World Energy Assessment. Energy and the Challenge of Sustainability.* New York: UNDP.

Urcan A.C., Marghitas L.A., Dezmirean D.S., Bobis O., Bonta V., Muresan C.I., Margaoan R. Chemical composition and biological activities of beebread – review. Bulletin of University of

Agricultural Sciences and Veterinary Medicine Cluj-Napoca. *Animal Science and Biotechnologies* 2017, 74(1), 6. https://doi.org/10.15835/buasvmcn-asb:12646

Urząd Rejestracji Produktów Leczniczych, Medycznych i Produktów Biobójczych. Farmakopea Polska VI [Polish Pharmacopoeia VI]; Urząd Rejestracji Produktów Leczniczych, Medycznych i Produktów Biobójczych: Warszawa, Poland, 2002; p. 5023.

USDA. 2019. *Agroforestry Strategic Framework. Fiscal Years 2019–2024*. Washington: U.S. Department of Agriculture.

USDA. 2022. *USDA Organic*, https://www.usda.gov/topics/organic [19.02.2023].

van Asselt, E.D., Arrizabalaga-Larrañaga, A., Focker, M., Berendsen, B.J.A., van de Schans M. G.M., van der Fels-Klerx H.J. Chemical food safety hazards in circular food systems: A review, *Critical Reviews in Food Science and Nutrition* 2022. DOI:10.1080/10408398.2022. 2078784

Vasić, V., Gašić, U., Stanković, D., Lušić, D., Vukić-Lušić, D., Milojković-Opsenica, D., Tešić, Ž., Trifković, J. Towards better quality criteria of European honeydew honey: Phenolic profile and antioxidant capacity, *Food Chem.* 2019, 274, 629–641. https://doi.org/10.1016/j. foodchem.2018.09.045.

Vermeulen S.J., Challinor A.J., Thornton P.K., Campbell B.M., Eriyagama N., Vervoort J.M., Kinyangi, J., Jarvis A., Läderach P., Ramirez-Villegas J. *et al.* Addressing uncertainty in adaptation planning for agriculture. *Proc. Natl. Acad. Sci.* 2013, 110, 8357–8362. https://doi. org/10.1073/pnas.1219441110

Vieux F., Soler L-G., Touazi D. *et al.* High nutritional quality is not associated with low green-house gas emissions in self-selected diets of French adults. *Am. J. Clin. Nutr.* 2013, 97(3). doi:10.3945/ajcn.112.035105

Vigar V., Myers S., Oliver C., Arellano J., Robinson S., Leifert C. A systematic review of organic versus conventional food consumption: Is there a measurable benefit on human health? *Nutrients* 2019, 12(1), 7. doi: 10.3390/nu12010007.

Visciano P., Schirone M., Berti M., Milandri A., Tofalo R., Suzzi G. Marine biotoxins: Occurrence, toxicity, regulatory limits and reference methods. *Front. Microbiol.* 2016, 7:1051. doi:10.3389/fmicb.2016.01051

Vlosky, R., Smithhart, R., 2011. A brief global perspective on biomass for bioenergy and biofuels. *Journal of Tropical Forestry and Environment* 1(1): 1–13.

Wan L-J, Tian Y, He M, Zheng Y-Q, Lyu Q, Xie R-J, Ma Y-Y, Deng L, Yi S-L. Effects of chemical fertilizer combined with organic fertilizer application on soil properties, citrus growth physiology, and yield. *Agriculture.* 2021; 11(12):1207. https://doi.org/10.3390/agriculture 11121207

Wang J., Li Q.X. Chemical composition, characterization, and differentiation of honey botanical and geographical origins. *Adv. Food Nutr. Res.* 2011, 62, 89–137. doi:10.1016/B978-0-12-385989-1.00003-X.

Wang R., Zou R., Liu J., Liu L., Hu Y. Spatial distribution of soil nutrients in farmland in a Hilly region of the Pearl River Delta in China based on geostatistics and the inverse dis-tance weighting method. *Agriculture* 2021, 11, 50. https://doi.org/10.3390/agriculture 11010050

WCED. 1987. Our Common Future. *The Report of the World Commission on Environment and Development*. New York: Oxford University Press.

Wehbe R., Frangieh J., Rima M., Obeid D.E., Sabatier J.M., Fajloun Z. Bee venom: Overwiev of main compounds and bioactivities for therapeutic interests. *Molecules* 2019, 24(16), 2997. https://doi.org/10.3390/molecules24162997

Weis W.A., Ripari N., Lopes Conte F., da Silva Honorio M., Sartori A.A., Matucci R.H., Sforcin J.M. An overview about apitherapy and its clinical applications, *Phytomedicine Plus* 2022, 2, 2, 100239. https://doi.org/10.1016/j.phyplu.2022.100239.

Weselek, A., Ehmann, A., Zikeli, S. *et al.* 2019. Agrophotovoltaic systems: Applications, challenges, and opportunities. A review. *Agron. Sustain. Dev.* 39: 35.

Weymann S. Autonomous agricultural vehicles – looking for new solutions. *Technika Rolnicza, Ogrodnicza i Leśna* 2017, 6, 4–8.

Wilde J. *Encyklopedia Pszczelarska. [Beekeeping Encyclopedia].* PWRiL, Warszawa, 2013. [in Polish]

Williams Ch.M. Nutritional quality of organic food: Shades of grey or shades of green? *Proc. Nutr. Soc.* 2002, 61, 19–24. DOI:10.1079/PNS2001126

Witczak A, Abdel-Gawad H. Assessment of health risk from organochlorine pesticides residues in high-fat spreadable foods produced in Poland. *J. Environ. Sci. Health* B *2014, 49, 917–28. doi:10.1080/03601234.2014.951574*

Wójcicki J. The effect of pollen extracts on the endocrine function in rabbits. *Herba Polonica* 1999, 27, 151.

Wojtacka J. Propolis contra pharmacological interventions in bees. *Molecules* 2022, *27*, 4914. https://doi.org/10.3390/molecules27154914

Woodward, L. 1996. Organic farming. In: R. Jones, M. Summers & E. Mayo (eds), *Sustainable Agriculture.* London: The New Economic Foundation.

World Health Organization. *The WHO Recommended Classification of Pesticides by Hazard and Guidelines to Classification 2009.* WHO Press. Geneva, 2010. Retrieved from http://www.who.int/ipcs/publications/pesticides_hazard_2009.pdf?ua=1

Wu J.G., Luan T.G., Lan C.Y., Lo W.H., Chan G.Y.S. Efficacy evaluation of low-concentration of ozonated water in removal of residual diazinon, parathion, methyl-parathion and cypermethrin on vegetable. *J. Food Eng.* 2007, 79, 803–809.

Yamada Y. Importance of codex maximum residue limits for pesticides for the health of consumers and international trade. In: Food safety assessment of pesticide residues Europe: *World Sci.* 2017, 269–282. doi:10.1142/9781786341693_0007

Yang Y., Luan W., Xue Y. Sustainability and environmental inequality: Effects of animal husbandry pollution in China. *Sustainability* 2019, 11, 4576. https://doi.org/10.3390/su11174576

You Ch.E., Moon S. H., Lee K. H., Kim K.K., Park W.C., Seo S. J, Cho H.S. Effects of emollient containig bee venom on atopic dermatitis: A double – blinded, randomized, base-controlled, multicenter study of 136 patients. *Annals of Dermatology* 2016, 28, 5, 593–599.

Young, J.Z. 1971. *The Study of Man.* Oxford: Oxford University Press.

Yu X., Dai Y., Zhao Y., Qi S., Liu L., Lu L., Luo Q., Zhang Z. Melittin – Lipid Nanoparticles Target to lymph nodes and elicit a systemic anti – tumor immune response. *Nature Communications 2020, 11, 1110. https://doi.org/10.1038/s41467-020-14906-9*

Żak N., Wilczyńska A., Przybyłowski P. Quality of foreign types of honey in comparison with Polish standards – preliminary research. *Problemy Higieny i Epidemiologii* 2017, 98, 3, 245–249.

Zaman M.M., Rahman M.A., Chowdhury T., Chowdhury M.A.H. Effects of combined application of chemical fertilizer and vermicompost on soil fertility, leaf yield and stevioside content of stevia. *J. Bangladesh Agric. Univ. 2018, 16(1), 73–81. doi:10.3329/jbau.v16i1.36484*

Zetterberg, L.; Filip Johnsson, F.; Elkerbout, M. 2022. *Impacts of the Russian Invasion of Ukraine on the Planned Green Transformation in Europe.* Stockholm: Mistra Carbon Exit.

Zhang A-A, Sutar P.P, Bian Q., Fang X-M, Ni J-B, Xiao H-W, Pesticide residue elimination for fruits and vegetables: The mechanisms, applications, and future trends of thermal and non-thermal technologies, *Journal of Future Foods* 2022, 2(3), 223–240. https://doi.org/10.1016/j.jfutfo.2022.06.004.

Zhang K., Dong R., Hu X., Ren, C., Li Y. Oat-based foods: Chemical constituents, glycemic index, and the effect of processing. *Foods* 2021, 10, 1304. https://doi.org/10.3390/foods10061304

Zhang L., Zhang Z., Sun D. Methods for measuring water activity (aw) of foods and ts applications to moisture sorption isotherm studies. *Crit. Rev. Food Sci. Nutr.* 2015, 57(5), 1052–1058. doi: 10.1080/10408398.2015.1108282.

Zhang Q.C., Shamsi I.H., Xu D.T., Wang G.H., Lin X.Y., Jilani G. *et al.* Chemical fertilizer and organic manure inputs in soil exhibit a vice versa pattern of microbial community structure. *Appl. Soil Ecol.* 2012, 57, 1–8.

Zhanggui Q., Xiaoping Y., Xia W. Trials of ozone reducing pesticide residues in grain. *Grain Storage* 2003, 32(3), 10–13.

Zikankuba V.L., Mwanyika G., Ntwenya J.E., Armachius J. Yildiz F. Pesticide regulations and their malpractice implications on food and environment safety. *Cogent Food & Agriculture* 2019, 5, 1, DOI: 10.1080/23311932.2019.1601544